Einstein's Dream

THE SEARCH FOR A UNIFIED
THEORY OF THE UNIVERSE

Einstein's Dream

THE SEARCH FOR A UNIFIED THEORY OF THE UNIVERSE

Barry Parker

PERSEUS PUBLISHING

Cambridge, Massachusetts

Library of Congress Cataloging in Publication Data

Parker, Barry R.
  Einstein's dream.

  Bibliography: p.
  Includes index.
  1. Unified field theories. 2. Cosmology. 3. Einstein, Albert, 1879–1955. I. Title.
QC173.7.P36   1986   530.1'42   86-15139
ISBN 0-306-42343-X

10   9   8   7   6   5   4   3   2   1

Perseus Publishing books are available at special discounts for bulk purchase in the U.S. by corporations, institutions, and other organizations. For more information, please contact the Special Markets Department at the Persues Books Group, 11 Cambridge Center, Cambridge MA 02142, or call (617)-252-5298

Preface

Thirty years ago Albert Einstein died, his dream of a theory that would unify the universe unfulfilled. He spent the last decades of his life searching for such a theory—a theory that would explain everything from elementary particles and their interactions to the overall structure of the universe. But he failed, not because he didn't try hard enough, but because the attempt was ahead of its time. When Einstein worked on the problem literally nothing was known about black holes, white holes, singularities, the Big Bang explosion and the early universe, quarks, gauge invariance, and weak and strong nuclear forces. Today we know that all these things are important in relation to a unified theory, and that they must be incorporated in and explained by such a theory. Thus, in a sense, our problem is much more complex today than it was in Einstein's day. But scientists have persevered and as a result we are now tantalizingly close to achieving this long-sought goal. Important breakthroughs have been made. In this book we will look at these breakthroughs and at recent unified theories—theories that go by the names supergravity, superstrings, GUTs, and twistor theory. In order to understand the problem, however, we must begin at the beginning.

In the first part of the book we will consider Einstein's general theory of relativity and the overall universe—in short, the macrocosm. From there we will go to the microcosm or world of particles and look at the important breakthroughs that have been made in this area. Finally, in the last part of the book, we will turn to recent unified field theories.

Although mathematics has been completely avoided in this book, it is impossible to avoid large numbers. Rather than write these numbers out in detail I have used what is called scientific notation. In this notation the number 10,000, for example, is written as 10^4 (the index gives the number of zeros after the one). For small numbers a negative sign is used before the index. The number 1/1000, for example, is written as 10^{-3}.

A brief comment on temperature scales is also perhaps in order. I have used what is known as the Kelvin scale (abbreviated as K). On this scale the lowest temperature in the universe is zero (it corresponds to −459 degrees Fahrenheit). The boiling point of water in this system is 373°K.

Finally, I would like to thank Linda Greenspan Regan for her careful editing of the text and many useful suggestions. I would also like to thank Sandra Carnahan for an excellent job on the line drawings.

Barry Parker

Contents

CHAPTER 1

The Goal

Looking into the dark night sky we feel a tingle of excitement as we are overcome by its grandeur and beauty. Each point of light we see is the image of a star, an image of light that may have left the star long before we were born. The universe is vast beyond imagination—almost terrifying in its intensity and complexity. The light from some objects you can see has traveled through space for over a million years, yet this same beam of light would cover the distance from the earth to the moon in less than 2 seconds.

Within this vast void our earth is but a speck, one of nine planets whirling around an insignificant yellow star we call the sun. Yet we are unique: the only planet in the solar system that supports intelligent life. But we cannot help wonder as we look at the stars—each a possible sun: Does any of them support life?

Our sun is one of about 200 billion stars in an island universe of stars—a galaxy—that we call the Milky Way. When the sky is sufficiently black, it can be seen as a faint silvery ribbon stretching from horizon to horizon. If we could somehow move out through these stars, beyond them to a point outside our galaxy, it would look like a fuzzy disk with a bulging center, its long arms wrapped around it in spirals. Our sun is located in one of these arms, about three-fifths the way from the center.

Most of the stars in our galaxy are ordinary like our sun, but a few are nothing short of amazing. Some pulse slowly as waves deep inside them surge outward forcing their surfaces to expand

Wide-angle view of the Milky Way in the direction of its center. (© 1978 AURA, Inc. Courtesy National Optical Astronomy Observatories.)

briefly, then subside. Others pulse so rapidly they would appear to our eyes to shine continuously. Still others—supernovae—explode in a dazzling display of power, their brightness increasing dramatically in a few hours as they spew out long tendrils of gas. We see the remnants of such an explosion in the Crab nebula—a star that exploded (as seen from earth) over 900 years ago, and is still expanding.

The remnants of other old supernovae—clouds of gas called nebulae—can be seen in many sections of the sky. Over millions of years gravity will pull these clouds, now enriched in heavy atoms as a result of the explosion, into new stars. Some of these new stars will, in turn, millions of years from now, also explode, creating another new generation of stars. This is the cycle of the universe: old stars exploding, new ones being formed from the debris. Millions of years will pass between generations, times that seem to us to be an eternity, but the universe (in its present form) is not eternal; it will not last forever; it had a beginning and it will, eventually, come to an end. According to present theories, it was born about 20 billion years ago in a gigantic explosion. When, or how, it will end we are still not certain. But it will end.

As mentioned earlier, the images of distant objects tell us only what these objects were like when the light we are seeing left them. The farther away the object, the older the image is. This means that when we look out into space with a telescope we are actually looking back in time. The galaxies we see tonight look as they did many millions of years ago. In the early 1920s Edwin Hubble of Mt. Wilson Observatory began studying these galaxies, and within a few years he had found something that both confused and astounded astronomers. They were all (with the exception of a few in our local group) moving away from us; in fact, the farther away they were, the faster they were traveling. The universe was expanding!

As we look farther and farther into space we eventually find that many of the galaxies are not like ours; they are in a state of turmoil. Immense forces are sweeping through them, tearing their stars apart as they throw them outward, generating in the process surges of radio waves that we can detect from here on earth. They are, in effect, exploding; we refer to them as radio galaxies. Probing even farther we come to objects that, despite 20 years of intense study, are still enigmas: the quasars. They are exceedingly energetic, pouring radio waves into space as if they were gigantic radio galaxies, yet from all indications they are tiny compared to galaxies—perhaps no bigger than supergiant stars. Indeed, they are so small and so far away that it is strange that we are able to see them at all.

Just beyond the quasars is the edge of our observable universe. It might seem odd that our universe has a boundary

The Andromeda galaxy, a spiral galaxy similar to our Milky Way. (© 1973 AURA, Inc. Courtesy National Optical Astronomy Observatories.)

because it immediately brings to mind the question: What is on the other side of it? To see why there is a boundary, consider the recessional speed of these objects. It is close to the speed of light, but according to the theory of relativity (which we will look at later), nothing can travel at the speed of light (usually designated as "c") relative to us. Just beyond these quasars, then, there is a point where, if there were objects here, they would have to be traveling at c—yet they cannot. This marks the edge of our observable universe.

Astronomers have learned a lot about the universe in the past few decades; many new types of objects have been discovered: pulsars, quasars. Ultraexotic objects such as black holes have been predicted to exist, and astronomers believe they have strong evidence that they do (but no irrefutable proof as yet). Each breakthrough, each new discovery, increases our understanding and knowledge of the universe, yet each also seems to bring with it new mysteries, new puzzles to solve, making us wonder, at times, if there will ever be an end.

So far I have been talking only about large-scale mysteries— mysteries associated with the universe, the macrocosm. But equidistant from us in the other direction is another frontier: the microcosm, or world of atoms and elementary particles. And there are equally as many mysteries here.

The universe is made up of many different types of particles, but of these three predominate: the electron, the proton, and the neutron. The electron has a negative charge; the proton has an equal but positive charge and is considerably heavier than the electron; the neutron has no charge.

The first of another group of particles—an entirely different type—was predicted in 1932 by the British theorist Paul Dirac. In working out a theory of the electron, he noticed that a particle like the electron in all respects except that it would have opposite charge, should exist. A few years later the positive electron, or positron, was found. It was eventually discovered that corresponding to all particles there are antiparticles, and a particularly strange phenomenon occurs when a particle and an antiparticle are brought together—they annihilate one another (with the release of considerable energy).

The year 1935 saw another important prediction. The Japanese physicist Hideki Yukawa postulated the existence of a particle with mass between that of the electron and the proton—a medium weight particle, or meson. And within 3 years a particle of medium mass was discovered, but it turned out that this particle, now called the muon, did not have the properties Yukawa predicted. In time, however, Yukawa's particle (now called the π meson, or pion) was found.

As the years passed, more and more new particles were found. At first they came in a trickle, but as accelerators got larger and larger the trickle became a flood, until finally there were so many "elementary particles" physicists felt overwhelmed. They began to wonder if an end would ever be reached. The particles were eventually grouped into two classes: the leptons and the hadrons. The leptons were light—the electron being the best known of this group—and the hadrons heavy. The hadrons were subdivided into two other groups called the baryons and the mesons. The best known baryon is the proton but the neutron also belongs to this family. The mesons, as mentioned earlier, are of intermediate mass.

But merely grouping these particles into families did not solve the problem. As the families grew, physicists soon realized there must be an underlying scheme; all of these particles—particularly the huge family of baryons—could not be "truly elementary." They had to be made up of other more fundamental particles.

In 1964 Murray Gell-Mann of Cal Tech and, independently, George Zweig in Geneva, Switzerland, suggested a way out of the dilemma. According to their idea, hadrons were made up of three basic particles that Gell-Mann called quarks (antiquarks also exist within the scheme). The theory was beautiful from a physicist's point of view—it accounted for all the observed particles, and reduced the number of truly elementary hadron types in the universe to a much more manageable three. But there was a problem—quarks had never been observed. No one had ever seen the track of an isolated quark in a bubble chamber; indeed, there was no evidence whatsoever that they existed, except for this theory. And despite the fact that they have not been found since, the theory has remained. It has been modified but it is still the best theory of elementary particles we have.

In short, then, the elementary particles of the universe, the ultimate bits of matter (as far as we know) can be grouped into two classes: the leptons and the quarks. We cannot split a lepton into anything more elementary, and of course we have not even isolated a quark, so we obviously cannot split it. It is now generally believed, in fact, that it is impossible to isolate quarks.

The world is made up, one way or another, of these particles. But, of course, if that were all there was to it the world would be a pretty strange place: countless billions of particles all drifting aimlessly in space. We know that these particles are not drifting aimlessly, but are held together by forces. There are, in fact, four basic forces of nature, two of which are particularly important in the atom. The atom consists of protons and neutrons closely packed together in a nucleus (almost all of the mass of the atom is here) that is orbited by electrons. In the electrically neutral atom the number of electrons is equal to the number of protons. The protons have a positive charge and the electrons have a negative charge, with the electrons being held in place as a result of the attraction (the electromagnetic force) between the charges.

If we look closely at the nucleus, though, we see that protons are held next to other protons. But protons have a positive charge, and like charges repel. A proton should therefore repel nearby protons, and indeed it does. There is another force, though, the strong nuclear force, that is about 1000 times as strong as the electromagnetic force, but differs from it in an important respect: it is short ranged, falling off sharply over any distance greater than the diameter of the nucleus. This means if we take two protons and bring them together, they will repel one another at first, but suddenly, when they are very close to one another there will be an extremely strong attraction, and they will be pulled together. The nuclear force does not act, however, between all particles, only between pairs of hadrons.

The third basic force of nature has almost no effect inside the atom; it is far too weak (over a billion billion billion times weaker than the electromagnetic force), but it is no doubt the one you are most familiar with—the gravitational field. Like the electromagnetic field, the gravitational field is long ranged, but it differs in an important respect: it is only attractive. (The electromagnetic field gives rise to both attraction and repulsion.) There is, of course, a small gravitational attraction between the nucleus and the electrons whirling around it in the atom, but it is so small it is completely negligible compared to the other forces. This does not mean the gravitational field is unimportant; we know it is important—it keeps us from falling off the earth. Furthermore, it keeps the earth in orbit around the sun.

The last basic force of nature is called the weak nuclear force. It is not as weak as the gravitational, but much weaker than either the electromagnetic or the strong nuclear. Like the strong nuclear it is extremely short ranged, but unlike the strong nuclear it is not an extremely common force, and occurs only when certain nuclear reactions take place.

THE SEARCH FOR MEANING

The modern scientific method—the idea of going into the laboratory and performing an experiment—was introduced by Galileo. Through it he was able to explain many phenomena of nature that had mystified humans for centuries. Later Newton introduced mathematics into science. He showed us that bodies in motion could be described by formulas, and that these formulas were not just a convenient shorthand, but had a certain magic. Not only could they tell us how a particle or body acted or moved in the past, assuming we knew the forces that it experienced, but they could tell us how it would act far into the future—for all time if need be.

Newton's most important contribution, however, was his introduction of the idea of a theory. Central to a theory were a number of basic laws from which numerous predictions could be made. Newton's theory of motion, or Newtonian mechanics as it is now known, was based on a few simple laws, and from them we could make predictions about any type of motion.

Soon after Newton introduced his theories, others emerged; the ideas of electricity and magnetism, over many years, finally gelled, with the help of Maxwell, into the electromagnetic theory. During these same years a theory of heat was formulated. All these theories, taken together, are now referred to as the *classical theory*.

The classical theory, for its time, was an excellent theory. It told us much about the universe—almost everything about the everyday universe of our senses. Embodied in it were relationships that had been verified again and again, and there was a basic simplicity about it. This, as far as most scientists are concerned, is important; a theory should be based on only a few postulates—the fewer the better. Furthermore, the theory should allow one to make verifiable predictions, and of course, classical theory did.

But all theories are, of course, man-made, and therefore suffer the shortcomings of their inventors. A new theory may seem, at first, to be a significant breakthrough, only to be discarded later. All theories are limited in their scope; if repeated experiments have verified their validity in a certain range, we can feel comfortable using them within this range. But we must be careful extending them beyond it.

Classical theory is such a case. Ordinary-sized objects at everyday speeds seemed to obey the classical laws of motion satisfactorily, but when scientists began to push the theory into the realm of the atom, and into the macroworld, they found it did not work. Something was wrong.

The faith in classical theory was so great, though, that it

took some time before we could accept its limitations. Several of these limitations were well known late in the 1800s, but most scientists saw them as only slight defects—things that could easily be patched up with a little work. At least one scientist even stated publicly near the turn of the century that almost everything about the universe (all the basic laws) was known. Little did he know that a revolution in physics was about to begin.

The first step in this new revolution was taken by the German scientist Max Planck. Planck was struggling with one of the shortcomings of classical theory in 1900 when he came to the realization that an entirely new approach was needed. He decided to test the idea that radiation (e.g., light) was emitted in "chunks" rather than continuously, as had previously been done, and lo and behold the idea worked. Although he thought he was merely patching up one of the classical equations, his "chunks," or quanta as he called them, were in reality an overwhelmingly important new concept. These quanta would soon become central to our understanding of the microworld.

A few years earlier, however, light had been shown to be a wave. How could it also be a particle (a quantum)? The difficulty was compounded in 1923 when a French prince, Louis de Broglie, proposed that the wave–particle duality that existed for light also existed for matter. He suggested that phenomena involving the interaction of radiation and electrons could best be understood if electrons acted as particles and waves.

The idea was thought at first to be the incoherent babblings of a madman. How could electrons possibly be waves? But de Broglie was, after all, of noble lineage and his Ph.D. thesis, which contained the proposal, could not be openly laughed or sneered at. But if it was accepted, and later turned out to be a horrible joke, the examining committee would be thoroughly embarrassed. It almost seemed as if they could not accept or reject it, so they called in an expert—Albert Einstein. To their surprise, Einstein was fascinated with the proposal, and told them he was convinced the idea was sound.

And sound it was, for in 1927, Davisson and Germer, in the United States, showed experimentally that electrons do exhibit wave characteristics. By directing a beam of electrons at a crystal, they produced a pattern of dark and bright lines on a screen; such a pattern would be produced only if the electrons were waves. Later it was shown that any type of particle produced a similar pattern; matter did, indeed, behave like waves.

The merging of the particle and wave ideas into a coherent mathematical theory was brought about in 1926 by Erwin Schrodinger and, independently, by Werner Heisenberg. But the theory they derived was strangely different—it was a probability theory. It did not give definite or exact predictions—only probabilities. We are, for example, given the results of such a theory every long weekend; on TV and radio come the predictions that 700 people will be killed over the weekend. And at the end of the weekend approximately 700 people are dead. Such projections can never, of course, tell exactly who is going to die. In the same way quantum theory can tell us that three out of ten atoms are going to radioactively disintegrate in the next 10 minutes, but it can never tell us which three.

Einstein made important early contributions to quantum mechanics but he was never convinced it was the final word. He felt it was at best an approximation; he believed there was an underlying theory that would eventually replace it, just as quantum theory replaced classical theory in the microworld. It was not that the theory gave inaccurate numbers; Einstein was satisfied with this aspect of the theory. It was its philosophical implications—what the theory told us about the physical world—that bothered him. Only probabilities could be calculated, not certainties; it was a statistical theory. It can tell us what will happen, on the average, to a beam of particles, but it cannot tell us what will happen to an individual particle in the beam. Einstein was sure there was an underlying theory that would allow us to determine what happens to individual particles.

Neils Bohr, the leading spokesman for quantum theory, became a close friend of Einstein, but he never shared his views.

Their views were, in fact, diametrically opposite and their debates over the philosophical implications of the theory extended over many years. Bohr's position was a little obscure and difficult to pin down but it eventually became known as the Copenhagen interpretation. It was based on a principle put forward by the German physicist Werner Heisenberg, now called Heisenberg's uncertainty principle, that tells us there is a "fuzziness" at the atomic level, and a principle that Bohr referred to as complementarity that has to do with the way we see elementary particles. An electron, for example, sometimes acts like a particle and sometimes like a wave. The principle says that these two aspects complement one another, which means that only one can exist at a time.

One of the questions that arose as a result of the Copenhagen interpretation was: "What do we mean by reality?" Quantum mechanics seemed to give an answer that made little sense, or at least was outside of what we usually refer to as common sense. Most of us believe that an objective world exists outside of us—in other words, a world is there regardless of whether we detect it or not. But according to the Copenhagen interpretation, this is not the case: what exists in the physical world depends on how we measure it. In fact, it does not exist before we measure it. An electron, for example, can be a wave or a particle depending on how we measure it. Furthermore, its position and momentum (the mass of a particle multiplied by its speed) depend on how we measure them.

Let us look at this last point a little closer—it stems from the uncertainty principle. According to this principle we cannot simultaneously measure both the momentum and the position of a particle. If we measure the momentum we disturb the position and it no longer has the position it once had. We might ask, then, if both position and momentum are really present. Potentially, they are, but each only becomes a reality when we measure it and since we can only measure one at a time we have to say the other does not exist. In short, there is no objective reality outside ourselves—it exists only when we take a measurement.

Further insight into this can be gained by looking back at the complementarity principle. An experiment that brings out several of its implications is known as the double slit experiment. Let us assume we are dealing with electrons. Suppose we have a beam of electrons striking a slit, and beyond the slit we set up a screen so we can see how the slit affects them. When there is a single slit, most of the electrons pass through as if they were particles; a few may be deflected slightly by the edges, but we will neglect them for now. The pattern we get is shown below:

The light pattern (top) from a single slit (left). The light pattern from a double slit (right). The height of the humps above the straight line is a measure of the amount of light that appears on the screen.

Now suppose that at some distance off to the side we introduce a second slit, and again let the electrons pass through. This time, strangely, we get an entirely different pattern on the screen.

To test things further, suppose we reduce the intensity of the beam until finally we have electrons going through the system one at a time. In this case we are certain each electron will go through either slit A or slit B. When we perform the experiment, though, we find that things are no different; if we run it long enough, the pattern on the screen is the same as it is in the double slit experiment above. But how could this be? For this to be possible, the electron that went through one of the slits would have to know if the other was open or closed. If it was open it would move to one position on the screen and if it was closed it would move to another. But how could it possibly know if the other slit is open? To answer this, we have to assume the electron is a wave that somehow expands out just before it enters the system and checks on the status of the second slit—in essence, it goes through both slits. Yet, being a particle, it seems physically that it must go through one slit or the other.

Perhaps we can outsmart the system. Suppose we set up an apparatus next to the slits and find out which slit it really goes through. Now, though, the uncertainty principle comes into play: by measuring the system we disturb it and change it. If we try to determine if it is going through A, we disturb it so it goes through B.

It was the "weirdness" of experiments such as this that was repugnant to Einstein. Because he was so strongly against quantum mechanics he was severely criticized. But Einstein's arguments were not the idle rebuffs of somebody who barely knew what the theory was all about. He had an intense interest in it and was said to carry Dirac's book, *The Principles of Quantum Mechanics*, with him wherever he went. And his effort to show there were problems with the theory was intense. He once said, "You can hardly imagine how hard I tried to find a satisfactory mathematical treatment of quantum theory. So far without success."

But Einstein was not entirely unsuccessful. He pointed out what appeared to be a flaw in the theory in 1935, and his ideas are still being seriously considered today. Along with Boris Podolsky and Nathan Rosen he published a paper titled, "Can Quantum Mechanical Description of Physical Reality Be Considered Complete?"

We will give only a simplified version of the argument presented in this paper. Assume we have an atomic system consisting of two particles, one spinning in one direction (call it spin-up) and one in the other direction (call it spin-down). Now suppose that we somehow cause the two particles to separate, i.e., fly off in opposite directions. Suppose further that one flies out of the window and goes off into space, and that the other one is directed toward an apparatus we have in the lab that can measure the spin of the particle. We measure its spin and find it is spin-up; this means the other particle has spin-down. Thus, we have determined something about the other particle without ever taking a measurement. But this is in conflict with the Copenhagen interpretation; it says that an object does not exist until it is measured: there cannot even be a particle without a measurement.

Obviously a way was needed to test this paradox. And in the mid-1960s it came. John Bell of CERN (European Organization for Nuclear Research) came up with a method of testing the idea in 1964. It is now referred to as the Bell inequality.

The first checks on the theory were inconclusive: some seemed to support Einstein's view, others supported Bohr's view. But an experiment was performed in 1983 at the University of Paris by Allain Aspect that seemed to clinch things. Bell's inequality was violated. Bohr was correct.

Does this mean quantum mechanics is the final answer, and that there is no possibility of an underlying theory as Einstein wished? Although things look bleak, this is not necessarily the case—science sometimes takes strange twists.

Einstein was not the only one who felt uncomfortable with quantum theory. The originator of the quantum idea, Max Planck, never fully accepted the theory, and Schrodinger eventually became convinced it was not the final answer. Even as recently as 1979 Paul Dirac made the statement, "It seems clear that the present quantum mechanics is not in its final form. It might very well be that the new quantum mechanics will have determinism in the way Einstein wanted. . . . I think it is quite possible that Einstein will turn out to be correct."

The Princeton physicist David Bohm has spent many years seeking an underlying theory. He is firmly convinced there are hidden variables within the theory. But the majority of physicists do not believe this is the case. They are convinced that quantum mechanics is the final theory and most accept the Copenhagen interpretation of it.

Another of the fundamental theories of physics was published in 1905—the theory of relativity. It dealt, not with the world of atoms, but with the concepts of space, time, and mass (and also electric and magnetic fields). The scientists Young and Fresnel had shown some years earlier that light exhibited certain phenomena (called interference and diffraction) that were consistent only with waves. And of course if light was a wave, a medium was needed to propagate it. To see why, suppose you throw a stone into a pond; a wave moves outward from the point where the rock hits. If there were no water at that point, however, there would be no wave. Obviously a medium is needed to propagate a wave; in this case the medium is water. In the case of light there is no obvious medium present so physicists invented one, calling it the ether. This ether presumably filled the entire universe, and its properties (transparency, incompressibility, not affected by gravity) made it difficult to detect. And although it alleviated the problem of how light waves were propagated, it created a new problem: it was a frame of reference for the universe. This meant that we could use it to determine our absolute velocity with respect to the universe as a whole. The ether was, in a sense, like a gigantic lake; we know we can easily determine how fast we are traveling across a lake (assuming we do not have a speedometer)—we merely set out a buoy and watch how fast it recedes from us.

To find out how fast the earth was moving through the ether, two physicists, Michelson and Morley, performed an ingenious experiment in 1887. They selected a beam of light as their buoy, projecting it in the direction of the earth's travel in its orbit. Since the ether was propagating the beam and the earth had a finite velocity in the same direction as the beam, it should catch up with it slightly. The beam should therefore appear to move away from us more slowly than it would if we were not moving after it (we are assuming the ether is generally calm around the earth). But when Michelson and Morley performed the experiment, they found, much to their surprise, that we (the earth) were not catching up. The light beam had a constant speed independent of ours; this implied that no matter how fast we chased after it we could never catch it. Light would always travel 186,000 miles/sec faster than us.

The scientific community struggled for years trying to make sense out of the enigmatic result. Appropriate formulas were derived independently by H. A. Lorentz in Germany and Fitzgerald in Ireland, but they did not explain what was going on. It was not until 1905 that a breakthrough came in the form of Einstein's special theory of relativity. Interestingly, Einstein had not heard of the Michelson–Morley experiment; he solved the problem as a result of looking at it from a different point of view. He was interested in what happened to electric and magnetic fields near the velocity of light, but the theory he devised encompassed much more than the effect on these fields: it told us how space, time, and mass were affected at speeds (relative to us) near the velocity of light: space is stretched (objects in it shrink), time slows down, and mass increases. (Actually, these effects occur at all velocities, but are only noticeable at velocities close to that of light.) It also told us that the concept of the ether was not needed.

In 1916 Einstein extended his special theory of relativity, which pertained only to uniform, straight-line motion, to include all types of motion. The result was the general theory of relativity (discussed in Chapter 2), which found that not only could space be stretched, but it could be twisted; indeed, it could be so grossly contorted that it no longer existed in our universe.

EINSTEIN'S DREAM

The two theories—the quantum theory and the theory of relativity—are still the two pillars of modern physics. One governs the microcosm, and the other (general relativity) the macrocosm, and both give excellent descriptions of their respective domains. Where classical (Newtonian) theory completely breaks down and is unable to give answers, these theories take over and give us highly accurate results. But the answers they give have, unfortunately, been obtained at the expense of abstraction. Where we could easily visualize what was going on when we applied classical (Newtonian) theory we no longer can. When we use these two theories we are forced to abandon the world of our perceptions; new and strange concepts are forced upon us.

But if classical theory breaks down when we deal with the microcosm and macrocosm, it is logical to ask if quantum theory and relativity also eventually break down. We saw earlier that as we go to higher and higher speeds we must supplement Newtonian theory with relativity theory. Quantum theory must be adjusted in the same way when we deal with high velocities. The resulting theory, called relativistic quantum mechanics, was devised by the English physicist Paul Dirac.

Looking again at our two pillars of physics—quantum theory and general relativity—we find that they are distinctly different theories. In essence, they speak two different languages and there seems to be no way they can communicate with one another. There is no bridge, no link, between them. But why should there be two such theories? Why not one that explains both the microcosm and the macrocosm. Furthermore, if we come back to the four basic fields of nature—the electromagnetic, the gravitational, and the strong and weak nuclear—we see another manifestation of the problem: the gravitational field is explained by general relativity while the other three fields are within the domain of quantum theory. No single theory explains all four fields. Furthermore, there are still difficulties with the elementary particles: how, for example, are the two fundamental families, the leptons and the quarks, related?

It was Einstein's dream that a single theory would cover all these phenomena—a unified field theory. Einstein's initial concern in this relation was quite modest; he merely wanted to unify the gravitational and electromagnetic fields, i.e., to find a single theory that explained both fields. He also hoped the theory would explain the nature of elementary particles. Unfortunately, he did not succeed.

This grandiose aim, a theory that unifies all physics, by bridging the gap between general relativity and quantum theory and explaining in a consistent, simple, unified way, all the fields of nature and their interactions with the particles, has not yet been realized, despite considerable effort over several decades. Einstein spent the last 30 years of his life looking for such a theory and other well-known physicists such as Heisenberg, Eddington, and Pauli also spent the last years of their lives searching for this seemingly unattainable goal.

But are we, in this search, really reaching for a rainbow? First, does it exist? And second, what if we found it? There would obviously be nothing else to learn about the universe if we did—a sad state of affairs as far as most physicists are concerned. Thus, we are caught in a kind of Catch-22; we search diligently for such a theory because it is in our nature to do so, but if we found it physics would suffer, as there would be no more goals to strive after.

Let us spend a few moments considering such a theory. Would it, in reality, have to explain everything? How far, in fact, can we know? Many physicists consider it naive to ask what we might call "ultimate questions." After all, we are still not even sure what an ultimate question is. A question such as: what is light?, may not seem ultimate but we cannot yet answer it. We know how light behaves, and we can describe this behavior with considerable precision, but we do not know exactly what light itself is. In fact, we do not know exactly what an electron is, or any particle for that matter. We can only describe their nature by probability functions.

It might seem from what was said earlier that there is an endless sequence of new theories, each improving on its predecessor. But do theories go on indefinitely in this way? It seems not, for inherent in quantum mechanics is a principle that does not allow this—the uncertainty principle. As you probe to smaller and smaller objects, a "fuzziness" sets in.

Does this mean that our present theories are as far as we can go? Definitely not, for as we saw earlier there are still numerous unanswered questions: the relationship between the four basic fields of nature, the connection between quantum theory and general relativity, the relationship between the quarks and the leptons, and the ultimate fate of the universe, to name only a few. We will continue striving for answers to these questions and for an ultimate unification, despite seemingly insurmountable barriers. We know our present theories give excellent descriptions of nature, but like their predecessors they are imperfect, and if we push things far enough they also break down. The conditions under which they break down, however, are far from our world of experience, even far from what we normally consider the microcosm and macrocosm. Before I talk about this breakdown, however, we must learn something about the theories themselves. I will begin with general relativity in the next chapter.

CHAPTER 2

Warped Space-Time

Before we dive into the deep dark waters of general relativity, we must learn something about space and time. You may feel you already understand these seemingly simple concepts: space is merely the void wherein everything exists, and time the chronological sequencing within this void. But there is actually much more to them than this. One of the first to take a penetrating look into their physical meaning was Isaac Newton. Born in England in 1642, Newton was sent to live with his grandmother when he was quite young. The genius that was later to reveal itself appears not to have been evident at this stage of his life. But he did have considerable mechanical ability; he loved to work with his hands, building windmills, water clocks, and other mechanical toys.

Like most great men of science, he had an uncanny ability for intense concentration. This was sometimes mistaken for absentmindedness, and it was occasionally embarrassing to him. For example, riding home one day, he came to a hill and decided to dismount and walk his horse up the hill. During the ascent his mind locked onto a problem and the world around him faded. When he got to the top he looked down at the bridle in his hand and suddenly realized the horse was not at the other end of it. He had been concentrating so hard he had not noticed the horse had slipped out of the bridle and wandered off.

Most people who knew the young Newton described him as a silent, sober, thinking person who preferred the company

Isaac Newton (1642–1727). (Courtesy AIP Niels Bohr Library.)

of girls to that of boys. Yet, strangely, he never married. He was humiliated and teased so much in early life that he later became suspicious of people and was extremely secretive about his work and ideas, always feeling that someone was trying to steal them. He was even reluctant at first to publish what eventually became his greatest work, the *Principia*.

Newton viewed space as absolute. "Absolute space, in its own nature, without reference to anything external, remains always similar and immovable," he wrote. And his influence was so great that few challenged his point of view. One of the few who did was the Irish philosopher George Berkeley. In his book *Principles of Human Knowledge,* Berkeley considered the problem of rotation in empty space, pointing out that if there was nothing else in the space it would be impossible to tell if an object (e.g., the earth) was rotating. We would not be able to detect its rotational motion unless there were other objects in the universe. He concluded on the basis of this that space was not absolute.

The German physicist Ernest Mach, famous for his book *The Science of Mechanics* and his stubborn views on atomic structure (he did not believe in atoms), picked up on Berkeley's ideas and extended them (they eventually became known as "Mach's principle"). To understand this principle, assume you are on a carousel that is spinning; you feel a force, and if you let go you know you will fly off. Mach considered the relationship of this force, called the centripetal force, to the rest of the universe. "What would happen to this force if all the other matter in the universe suddenly disappeared?" he asked. He convinced himself that it would also disappear. If there was nothing else in the universe, we would not know we were rotating—in fact, the concept would have no meaning. And if there was no rotation there would be no centripetal force. This meant that local forces had to depend not only on local things, but on the universe as a whole—even the most distant stars. When Mach put his ideas forward they were considered outrageous; people laughed. How could distant stars have anything to do with the force you feel when you spin?

But Einstein did not laugh; he was, in fact, fascinated by the idea and spent several years considering its consequences. He later referred to it as a guiding light in the formulation of his general theory of relativity. It turned out, however, that although Einstein had hoped to incorporate the principle directly into his theory, he eventually found out that it was not needed.

Einstein's theory showed us that space is not the rigid,

absolute void that Newton visualized. It is, in a sense, a physical "thing"—a thing much more complex than we ever dreamed. Not only can it be stretched, twisted, and changed from point to point, but we will see later that particles can suddenly appear in it—apparently created out of space itself. Indeed, we are still likely ignorant of many of its properties—properties we cannot even visualize today.

The second of our fundamental concepts is time, a concept even more enigmatic than space. We feel the passage of time, and can easily distinguish the instant we call "now" from an instant in the future or past, and consequently we think we understand it. But the physical time we experience is not the same as mathematical time. Einstein pointed out the fallacy in a humorous way when asked once about relativity. "When a pretty girl is sitting in your lap, an hour seems like a minute," he said. "But a minute sitting on a hot stove seems like an hour."

As in the case of space, Newton was convinced that time was also absolute—never changing, always the same at every point in the universe. We have already seen that special relativity tells us this is not so. If an astronaut takes off from earth and passes overhead at a velocity close to that of light, we notice a significant difference in the rate of his clock compared to ours. His appears to be going much slower, and the closer his velocity is to that of light, the slower it goes. Strangely, if he looks back at our clock he does not see, as you might expect, that it is speeded up; he also sees our clock running slow. Yet, when the two clocks are brought back together they will not show the same time. The astronaut's clock will show less elapsed time than the clock back on earth. How is this possible if both see each other's clocks run slow? The answer to this question comes only from general relativity, and it was one of the reasons Einstein felt compelled to extend special relativity—it was obviously incomplete.

The elasticity of time is used extensively in science fiction, but in most cases it is pushed beyond its limit: trips to the past and future are common in the genre. These trips were usually said to be a result of passage through a "time warp." People who pass through these time warps—which are frequently shown as gigantic whirlpools or cyclonic storms—suddenly find themselves in a world of the future or past. I have to admit that I have been as fascinated by such stories as anyone, and I do not want to shatter anybody's fantasies but, facing fact, we now know that such trips are impossible. There is a fundamental concept called *causality*—the idea that every event is caused by some other event—that forbids them. To understand causality we must look back again at what happens to time at high relative velocities, or perhaps I should say, at any velocities, since the phenomenon occurs at all relative velocities, but is only noticeable when the difference is large. Consider two observers A and B; assume A stays on earth and B moves relative to earth in a rocket ship. As B moves faster and faster, his clock, according to A, moves slower and slower until finally just before he is moving at the speed of light his clock almost stops. If we push things a little further, in theory, we would see that B's clock, as seen by A, would stop at the velocity of light. But in reality this cannot happen, because one of the fundamental results of special relativity is that matter cannot move at the velocity of light; it is, in fact, this unattainability that gives us causality. If we were able to travel at greater velocities we would be able to journey into the past and into the future. And, of course, if this were possible we would be able to tamper with them. You could, for example, kill your grandfather at a very young age, which would not allow "you" to exist—creating an obvious contradiction.

So, despite the intrigue that time warp stories generate, they are not scientifically sound. Of course, it might, in some strange and yet unknown way, someday be possible to observe the past without tampering with it. We certainly have no idea how this would be accomplished, but it is dangerous to commit ourselves to the impossibility of anything—we must keep an open mind.

Another concept related to time that we feel we understand

is the "flow" of time. Time is obviously a one-dimensional concept: we can relate to a past, a now, and a future. It may seem therefore that it is like a never-ending river that flows forever—the "river of time." But physicists do not see it this way; they point out that we cannot measure the flow of time. Clocks do not measure this flow; they merely measure intervals of time. We label these intervals with numbers, but these numbers are akin only to the mileage signs along a road. The speedometer in a car tells us how fast, or at what rate we are passing these road signs, but a clock does not tell us how fast we are passing various time intervals. A car can speed up and slow down, and the speedometer will tell us by how much. A clock does not do this. Thus, time, like space, is also much stranger than we might have imagined.

CURVED SPACE

Intertwined with the concept of space is a subject you likely have some familiarity with—namely, geometry. The earliest and best known geometry was formulated by the Greek mathematician Euclid. Although little is known of Euclid personally, his geometry book *Elements* is one of the most studied books in Western culture. Since its origin it has gone through more than 1000 editions. Your high school geometry was based on it.

Near the beginning of this little book are five axioms, or self-evident truths. The first four seem somehow more fundamental than the fifth. And through the centuries mathematicians have puzzled over this fifth axiom, wondering if it was truly an axiom, or perhaps a theorem that could be proved from the other axioms. It can be stated as follows: Assume we have a straight line and a point off to one side of the line. Through this point one and only one line parallel to the existing line can be drawn.

The first person to see a flaw in this apparent self-evident truth was the German mathematician Karl Gauss. What Gauss realized is that Euclid's geometry, in two dimensions, was a geometry of a flat surface. He considered the consequences of a transfer of this geometry to a curved surface (e.g., the surface of the earth) and noticed that the fifth axiom was no longer valid. To see why, consider a straight line on the earth, say a line of longitude; if you select a point nearby and try to draw a line through it parallel to the first line, you soon find you cannot. The straight lines on a sphere are great circles (e.g., lines of longitude). If you tried to draw a straight line parallel to an existing one, you would find they crossed—just as all lines of longitude cross at the poles on earth.

There are also other differences in curved space geometry. We know, for example, that if a triangle is drawn on a flat piece of paper, the sum of the interior angles is 180 degrees (two right angles). On the surface of a sphere this same triangle would contain more than 180 degrees, depending on how large it was compared to the radius of the sphere.

Gauss's ideas on non-Euclidean geometry were picked up and expanded upon by Georg Riemann, one of his students. Riemann suffered from frail health throughout much of his life and died at the age of 40, yet in this short time he literally wrote the book on non-Euclidean geometry. Where Gauss had considered his geometry only in two dimensions, Riemann generalized it to three and even more dimensions. It is easy for us to visualize a curved surface, but what exactly is a curved three-dimensional space? This was something entirely new to mathematics—something that could be dealt with using symbols and numbers, but not something that could be visualized. Riemann was undaunted; he did not care whether he could imagine it or not—he could make calculations and predictions, and that is all that mattered.

About the same time that Riemann was expanding our horizons on Gauss's developments, two other mathematicians, Nikolai Lobachevski and Janos Bolyai, were independently developing another non-Euclidean geometry. They were interested in what kind of geometry would result if an infinite number of lines could be drawn through the point beside the line. Bolyai developed the geometry and sent the results to his father, who passed them on to Gauss. Lobachevski published his results in a book on geometry. There were now three possible geometries, two of them based on a variation of the fifth axiom. In Euclid's geometry only one parallel line could be drawn through the point, in Riemann's none, and in the Lobachevski–Bolyai geometry an infinite number. Although each of these geometries applies to two, three, and even more dimensions, it is easiest to visualize them in two dimensions. As I mentioned earlier, Euclid's is a flat surface geometry and Riemann's a curved surface geometry, but the surface in the case of Riemann must be positively curved, like the surface of a sphere. The Lobachevski–Bolyai geometry can be visualized as that on a negatively curved surface; to us this surface would look like a saddle. If we drew a triangle on it, the interior angles would add up to less than 180 degrees.

In addition to the geometry discussed above, Riemann also generalized his geometry so it could be applied locally. This generalized geometry could account for variations of curvature from point to point. If the surface had hills and valleys, for example, they could be described. To see how this is accomplished, consider Pythagoras's theorem; you are likely familiar with it from school. According to this theorem, if a right angle triangle is drawn on a flat surface, the sum of the squares of its sides is equal to the square of the hypotenuse. If, however, the surface is not flat, this relationship is not satisfied—a new relationship that depends on the curvature of the surface will replace it. This means that if we measure the lengths of the sides of a right triangle, we can determine the curvature of a surface. And if the curvature varies from point to point, we need merely cover the surface with small triangles and measure their sides.

Riemann's work was extended by the mathematicians Ricci and Christoffel. The culmination of their work was an abstract but beautiful branch of mathematics called tensor analysis. This was the tool that Einstein would use in setting up his general theory of relativity.

We saw earlier that Riemann considered curved mathematical spaces of three and more dimensions. He did not, however, restrict his thoughts to "mathematical" spaces, but considered the possibility that the physical space of our experience might be curved. It is, of course, impossible for us to visualize this physical space; the best we can do is visualize a two-dimensional space (a surface) embedded in the three-dimensional space embedded in a higher-dimensional space mathematically. It might seem that four dimensions would be appropriate, but it turns out that four is not enough. To determine a three-dimensional geometry properly we actually need six dimensions.

Einstein is sometimes given the credit for being the first to consider the possibility of curved physical space, but he was not. In addition to Riemann, the mathematician William Clifford was an enthusiastic advocate of the idea: ". . . small portions of space are of a nature analogous to little hills on the average flat . . . the ordinary laws of geometry are not valid on them. . ." he wrote. But the idea was ahead of its time and was generally ignored by his contemporaries. I should point out, though, that while Clifford championed the idea of curved space, unlike Einstein he did not set up a mathematical theory showing how and why it was curved. What Clifford did and what Einstein did are on two different levels. Einstein realized, of course, that he owed people such as Riemann and Clifford a considerable debt. Without Riemann's advances, he once said, he would have been unable to formulate his theory.

EINSTEIN AND WARPED SPACE-TIME

Einstein, who would bring the ideas of curved space, gravity, and time together into a coherent and consistent theory, was born in Germany in 1879. His father was an easy-going, jovial, but not too successful small-factory owner, and his mother a sensitive, understanding person with a deep love of music. Although his parents worried about how late he had begun to talk, his grandparents believed he was outstanding already at an early age. His grandfather wrote, ". . . I just love that boy, because you cannot imagine how good and intelligent he has become."

Einstein admitted later than he did not like school; the teachers reminded him of drill sergeants. He was unlike other boys his age in that he did not like anything about the military: he refused to engage in children's war games and abhored military parades. But already at an early age he had a love of nature and a passionate curiosity about its workings. It is interesting to speculate on the things that may have sparked this curiosity. When asked about them he usually referred to two: the gift of a compass when he was about 5 and a geometry book at 11. That the compass always pointed in the same direction regardless of how he moved it seemed miraculous to him.

When Einstein was in gymnasium (equivalent to our high school) his father's business failed and the family moved to Italy. It was decided, however, to leave Einstein in Germany to finish his education. Left alone at a school that employed a dull, mechanical method of teaching, he felt forlorn and unhappy and was soon scheming to rid himself of it. To his surprise, however, his homeroom teacher called him in one day and asked him to leave. (His attitude had apparently become evident in the classroom.) Overjoyed, he left on foot for northern Italy where his parents were living, and for the next several months he vented his energies and adventuresome spirit on the nearby mountains and on books he enjoyed. It was a delightful, heady experience, one he long remembered. But once again his father's business began to fail and he was told to start thinking about a vocation.

He still did not have his gymnasium diploma and most universities would not have admitted him; however, the nearby Zurich Polytechnic required only that students pass an entrance exam, so Einstein decided to take it. He was 16 years old, about

2 years younger than most of the students taking the exam, but he had already taught himself calculus and had read extensively—mostly popular science books. Surprisingly, the germ of later ideas was already planted; as a result of his reading he had begun to wonder what it would be like to travel at the speed of light. "What would happen to a light beam if you traveled alongside it?" was a question that had already entered his mind.

But his expertise in mathematics and science did not help him on the exams—he failed. It is perhaps consoling to think that Einstein failed at something; it gives the rest of us hope, I suppose. Anyway, the problem was not serious. His knowledge was apparently not broad enough; he lacked a proper grounding in areas such as biology and languages. The examiner noticed his excellent showing in the math and science parts of the exam, however, and suggested that he get a high school diploma and try the next year.

Einstein followed his advice and enrolled in a school in Aarau, a small town not far away. He was delighted to find that Swiss schools were quite different from German ones: the stark militarism was replaced with a casual, friendly atmosphere that Einstein thoroughly enjoyed. The following year he passed the Zurich Polytechnic exam and was admitted.

But Einstein was not an ideal student—far from it. He skipped classes when his interest lapsed, or when the professor did not cover material he wanted to learn. The physics class was a disappointment to him because Maxwell's theory of electromagnetism was not covered. His math teacher told him he was "brilliant . . . but a lazy dog." The truth is, though, that when he was not in class he was in the laboratory performing experiments or studying the works of Maxwell, Helmholtz, and others on his own. Nevertheless, the day of reckoning came. Einstein had let many of his classes slide and thus had to cram extensively for final exams. Fortunately, he had a friend, Marcel Grossman, who was an excellent student and with the help of Grossman's notes he passed. But it was a bitter experience; he later looked back on the extensive cramming he had done with

Albert Einstein (1879–1955). (Courtesy AIP Niels Bohr Library.)

considerable distaste. "I couldn't even think of anything scientific for almost a year after I graduated," he said. It makes us wonder if force-feeding is the best way.

Although he now had a degree, he was in for yet another shock. He had hoped to become an assistant to one of his professors, but because of his aloofness in class he was an outcast—nobody wanted him. He spent several months writing resumés and being interviewed but generally getting nowhere. He worked for a while as a substitute teacher, but got into trouble for teaching material that was not required and was soon released. It was the low point of his life, a period of severe disappointment and disillusionment. Lesser men would likely have folded their cards and left physics but Einstein's love was by now too deeply rooted. Despite his rejection he wrote his first scientific paper during this time; it was published in the prestigious *Annalen der Physik* in 1901.

Finally, though, through the bleakness came something positive in the form of his old friend Marcel Grossman. Although Grossman had secured a position in mathematics at Zurich Polytechnic he was still a junior member and in no position to offer Einstein a job, but he had spoken to his father about Einstein and an interview was arranged at the patent office in Bern. And Einstein was hired. Curiously, this job, almost degrading for somebody with a teaching degree in physics, turned out to be to his advantage. Einstein enjoyed the work; he had always been interested in gadgets and machines and how they worked—in particular, on what scientific principle they were based. Some of these inventions no doubt influenced his early work on thermodynamics. But the best thing about the job was that it left him considerable spare time—time in which he could work on physics. He later referred to the patent office as "that secular cloister where I hatched my most beautiful ideas. . . ."

Shortly after he obtained the job he married Mileva Maric, a student he had met at Zurich. Although he was happy with family life and had considerable free time to pursue his interests, his wages were so minuscule he literally lived the life of a pauper. "If everyone lived as I do romantic novels would never have come into being," he once said. On another occasion, "In my theories I put a clock at every point of space, but in reality I can hardly afford one for my house."

But the atmosphere was good and Einstein's inventive powers were near their peak. Everything gelled and in 1905 he laid a golden egg: the theory of relativity. The merits of the theory were not recognized immediately, however; many years passed before it was generally accepted.

In setting up the theory Einstein still treated space and time separately. It was H. Minkowski, one of his teachers at Zurich—the one who had called him a lazy dog—who brought them together. He showed that a four-dimensional unification of them into what is now called space-time was a much more meaningful concept.

To see the significance of his unification, consider two astronauts somewhere out in space. Both witness the same explosive event, but because they are different distances from it, they see it at different times. One insists it occurred at 3:00 and the other that it occurred at 6:00. They are both right, of course, but we know the "event" itself had to occur at one particular time. If, instead of considering just time, we consider the event in space-time, it would have been the same for both observers.

The easiest way to represent space-time is on what we call a space-time diagram. Since we only have two dimensions on a page we let one of them represent the usual three dimensions of space, and the other, time. A point in this space-time diagram is an event. An example of an event is an astronaut somewhere in space pressing a button so that a flash of light is emitted; the

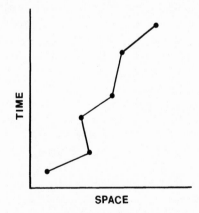

A space–time diagram.

flash or "event" occurs at a specific point in space and at a specific time. If he presses it again a few minutes later a new event occurs. Each single event is represented by a point in our diagram; actually, just *being* at a particular point in space at a particular time is an *event*. The sequence of these events is called a world line.

Intervals in space-time are the same for all observers, and are called *invariants*. Invariants, as we will see later, are particularly important in relativity. Einstein, in fact, would have preferred to have called his theory the theory of invariants rather than relativity theory, but the latter was forced upon him. He showed his distaste by referring to it as "the so-called relativity theory."

Soon after Einstein formulated his special theory of relativity he realized it had to be generalized. It applied to objects traveling with uniform speed in a straight line, but it did not apply to accelerated objects. Accelerated motion was different from uniform motion; if you were moving uniformly you could not tell you were moving unless you saw something moving relative to you, but you could tell if you were being accelerated; you would feel a force on your body. If you were in an accelerating car, for example, you would be pushed back into the seat; similarly, as it decelerated you would be thrown forward.

Einstein began an investigation into the difference between accelerated and uniform motion by considering the force we feel when we are accelerated; it is called an inertial force. To obtain an understanding of this force, consider several balls of different mass, ranging from small plastic ones to large metal ones, sitting on a table. If you give each a push, you know that it will take considerably more effort to move the larger ones than the smaller ones. Moreover, if you push each with an equal force, you will see that the acceleration imparted to the smaller ones is much greater than that imparted to the larger more massive ones; in fact, the more massive the ball, the less the acceleration. Yet if you took these same balls and dropped them, they would all accelerate to the ground at the same rate. This seemingly strange behavior occurs because there is a greater gravitational pull on the larger masses; there is, in fact, an exact balance between the gravitational pull and the force of inertia. This was not a discovery of Einstein's, having been known to Newton. Newton realized these forces were equal, and since there was a mass associated with each of them, these masses, called the "inertial mass" and the "gravitational mass," were also equal. Furthermore, it was generally assumed by Newton and his contemporaries that no mechanical experiment could distinguish them.

Einstein extended Newton's ideas in what was later called the principle of equivalence. He stated that no experiment whatsoever—mechanical or otherwise—could distinguish between inertial mass and gravitational mass. Scientists now refer to Einstein's statement as the strong equivalence principle and Newton's as the weak equivalence principle.

A more complete understanding of the principle comes from considering an observer in a windowless elevator. Assume this elevator is taken out into space (far from the gravitational influence of the earth) and pulled upward with an acceleration of 32 feet/sec^2 (the acceleration of gravity here on earth). The observer would immediately think he was on the surface of the earth: If he dropped something it would fall as it does on earth, and if he jumped he would jump as he does on earth. According to the equivalence principle, in fact, everything he did would be the same as if he were in the earth's gravitational field. There is no experiment he could perform that would distinguish between this field and the force generated by the accelerating elevator. The "inertial field" generated by the acceleration would be completely equivalent to a gravitational field. Thus, in attempting to generalize his theory of special relativity from uniform to accelerated motion, Einstein found that he was developing a theory of gravity.

Einstein was still in the patent office when he published his first paper on general relativity; it contained the principle of equivalence. He used this principle to show that accelerated motion was not absolute. Inertial forces, created by acceleration, cannot be distinguished from gravitational forces. They are equivalent. All motion, including accelerated motion, is therefore relative. One way of thinking of this is to consider an inertial pull as due to an acceleration relative to the rest of the universe, and a gravitational pull as due to an acceleration of the rest of the universe relative to us.

Einstein left the patent office in 1909 to take a position at the University of Zurich, but his stay was brief, for in 1911 he was offered a position at the University of Prague, which he accepted. In the same year he published a second paper on general relativity in which he considered some of the consequences of the principle of equivalence. One of the results of special relativity was the equivalence of mass and energy. Using this in conjunction with the equivalence principle, he showed that a light beam passing through the gravitational field of the sun would be deflected by about 0.83 sec of arc—an extremely small amount, but large enough to test. Unknown to Einstein, the same calculation had been done by the surveyor-mathematician Johann Soldner over 100 years earlier. In an attempt to discredit and embarrass Einstein, the German physicist Philipp Lenard, an enthusiastic Nazi supporter, had Soldner's paper republished in *Annalen der Physik* in 1921, but it had little effect.

Like his stay at the University of Zurich, Einstein's stay at Prague was brief. Grossman and others at his old alma mater, Zurich Polytechnic, were out to lure him back, and in 1912 he accepted their offer. With his second stay at Zurich came a new phase in his work on general relativity. He had by now formulated many of the physical ideas and had considerable insight into what direction the theory should take, but he did not have a mathematical structure. His move was a stroke of luck in that his old friend Grossman was an expert in the mathematical techniques he needed. Grossman introduced him to the works of Riemann, Christoffel, and Ricci—tensor analysis (a branch of mathematics dealing with tensors, which are mathematical quantities that have specific properties)—and Einstein realized immediately that this was exactly what he needed. He had earlier formulated a second principle called the principle of covariance that stated that the laws of physics should be independent of the coordinate system (a grid system used to denote positions in space-time), and tensor analysis allowed him to express this mathematically. Under Grossman's guidance, Einstein studied and learned tensor analysis. The two men actually published several papers together, but the work at this stage was still a kind of groping in the dark—a mind-tormenting search for the one equation out of hundreds that was the correct one. Numerous candidates were tested and rejected. Interestingly, the correct equation was actually considered briefly, but rejected by Einstein because he came to the mistaken conclusion that it violated causality. Then a further mistake led to the abandonment of the principle of covariance.

While Einstein was at Zurich he was visited by Planck and Nernst from the University of Berlin. They made him an offer he could not refuse: he would be director of the research branch of the newly planned Kaiser Wilhelm Institute. And in addition he would have almost total freedom to pursue his research (he would not have to teach). This aspect of the offer was particularly pleasing to him; it was not that he did not like contact with students—he did, but he found routine lecturing restricting. It was, indeed, the considerable free time that he had here that allowed him to finish the theory.

By the time Einstein began working on relativity in Berlin, his expertise in tensor analysis was finely tuned and soon everything began to come together. By mid-1915 he had decided that the only problem in regard to the principle of covariance was his previous stupidity and he reincorporated the principle into his theory. Some insight into his final struggle is brought out by his friend Charlie Chaplin in his book *My Autobiography*.

Chaplin recalls a dinner at his home in California in 1926, at which Einstein, Mrs. Einstein (his second wife), and two other friends of Chaplin were present. At dinner Mrs. Einstein "told [him] the story of the morning [Einstein] conceived the theory of relativity." She related:

The Doctor [Einstein] came down in his dressing gown as usual for breakfast but he hardly touched a thing. I thought something was wrong, so I asked him what was troubling him. "Darling," he said, "I have a wonderful idea." And after drinking his coffee, he went to the piano and started playing. Now and again he would stop, making a few notes then report: "I've got a wonderful idea, a marvelous idea." I said: "Then for goodness sake tell me what it is, don't keep me in suspense." He said: "It's difficult, I still have to work it out."

Mrs. Einstein told Mr. Chaplin that Einstein continued playing the piano and making notes for about half an hour, then went upstairs to his study, telling her that he did not wish to be disturbed, and he remained there for two weeks. "Each day I sent up his meals," she said, "and in the evening he would walk a little for exercise, then return to his work again."

"Eventually," Mrs. Einstein said, "he came down from his study looking very pale." "That's it," he told me, wearily putting two sheets of paper on the table. And that was his theory of relativity.

Einstein presented his theory at the next three consecutive sessions of the Prussian Academy of Sciences in November 1915. He later referred to this as the happiest time of his life.

Einstein had, from the beginning, disliked Newton's idea of gravity as an action-at-a-distance field. Gravity's influence, according to Newton, was instantaneous: if an apple suddenly dropped from a tree in front of you, the entire universe knew, and adjusted for the difference, immediately. But according to special relativity, nothing can travel at a velocity greater than that of light. Pondering the problem, Einstein thought about the beam of light curving around the limb of the sun; he soon realized that it was not the beam that was bent, but rather the space through which it traveled. Matter must somehow curve space and other matter must move through this curved space in the way we see it move—yet this way must be "natural." He decided the most natural way would be along a path that represented the shortest distance between two given points in space (this is called a geodesic in mathematics). This would mean that the sun curves the space around it and the planets move in this space along geodesics. These geodesics appear to us to be elliptical orbits, but in curved space they are actually straight lines.

Not everyone, by any means, accepted Einstein's strange new view. Someone was observed rushing out of one of his lectures shaking his head and muttering, "curved space . . . such nonsense . . . how could space possibly be curved . . . the man belongs in a nuthouse."

It was, of course, important that Einstein give his theory credibility. Just predicting curved space was certainly not enough. Besides, it was well known that Newton's theory itself was quite accurate. If Einstein's theory was better, it had to, as a first approximation, give the same results as Newton's theory, but in addition it would have to predict things that Newton's theory did not. If it only duplicated Newton's theory, it would not be of much value. Einstein showed that in the first approximation it did, indeed, give the same results as Newton's, and in its exact form it did predict more.

Before we consider these further predictions, let us consider his idea of curved space in more detail. The equation that he derived tells us exactly how much and in what way the space is curved around a given mass; it also tells us how the space is curved inside the mass. But, as mentioned earlier, we cannot visualize three-dimensional curved space, we can only visualize a two-dimensional surface in a three-dimensional space. The best way to think of this curved surface is as a thin rubber sheet

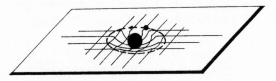

A simple representation of the curvature of space around the sun. The heavy ball at the center of the sheet represents the sun; the small ball represents the earth.

that has been stretched and tacked at the edges. We can place a large, heavy ball in the center of it to represent the sun. It will dent and curve the sheet around it just as the sun curves space. If we then take a marble and project it as shown in the figure on the preceding page, it will move around the ball several times in an elliptical orbit, just as the planets orbit the sun.

Now, let us consider the ideas in Einstein's theory that are not contained in Newton's theory. We will begin with the orbits of the planets. Newton's theory predicts that the planets should travel in ellipses, and that they should stay indefinitely in the same orbit. It had been noticed many years earlier, though, that Mercury did not appear to follow its predicted orbit faithfully. Many astronomers thought there was another planet between Mercury and the sun that was perturbing it; however, when Einstein calculated the orbit using his theory, he noticed that it gave slightly different results than Newton's. The equations were the same except for an additional term, but this additional term predicted a slow change in direction of the orbit's major axis—a precession. And when astronomers compared Einstein's prediction against observation, there was excellent agreement. The conclusion was inescapable: Einstein's theory was better. Actually, the theory predicts a precession of all planetary orbits, but the effect is only observable in the case of Mercury.

Another of Einstein's predictions was the deflection of a beam of light around the sun discussed earlier. In 1911 Einstein predicted a deflection of 0.83 sec of arc, based on a calculation using Newtonian mechanics. With his new theory he redid the calculation and found a deflection twice as great. Although he had encouraged astronomers to measure the deflection after his earlier prediction, this had not been done. An expedition set off to Russia to observe an eclipse in 1914, hoping to measure the deflection of the stars near the sun, but World War I began and they were taken prisoner. It is perhaps fortunate that the measurement was not completed. It would have turned out to be twice as large as predicted and considerably less attention would have been paid to Einstein's theory.

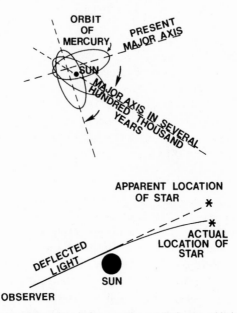

(Upper) Precession of the orbit of Mercury. (Lower) Deflection of light by the sun.

Verification of the second prediction came shortly after the war. Sir Arthur Eddington, in England, had become fascinated with Einstein's theory, and even before the war was over he was busy organizing an expedition to South America to observe the 1919 eclipse. He was almost as confident as Einstein was that the theory would turn out to be correct. It might seem strange that British astronomers were planning an expedition to verify a German's theory, while they were still at war with Germany. But both Einstein and Eddington were pacifists.

There were actually two separate expeditions to the eclipse. One expedition got a deflection of 1.61 and the other 1.98, the average of the two being very close to Einstein's prediction of 1.75 sec of arc. Einstein was jubilant when he got the news but his confidence showed even then. A student asked him what he would have done if the prediction had turned out to be wrong.

"Then I would have felt sorry for the dear Lord—the theory *is* correct," he replied.

The verification made Einstein a celebrity almost overnight. He was surprised and to some extent annoyed at the hoopla that the press made over the event. Reporters plagued him for months, then years. It was perhaps a lucky accident that the result was as close as it was to Einstein's prediction. Conditions on eclipse expeditions are far from ideal for highly accurate work. In some of the later expeditions, considerably different results were obtained—ranging from 1.18 to 2.24. This does not mean the prediction is wrong, just that on expeditions of this type it is extremely difficult to measure such small angles accurately. With more sophisticated equipment we can now observe the bending of electromagnetic waves around the sun (the effect applies to all electromagnetic waves, not just light) without the need of an eclipse. These experiments have verified Einstein's results to within about 1%.

A third prediction that came out of Einstein's theory was that time should run slower as we moved to stronger gravitational fields. The effect would be extremely small except when the difference was exceedingly large. We know in the case of the earth that as you move upwards above its surface the gravitational field decreases. This means that if we synchronized two clocks at its surface, then moved one, say 1000 feet up, it will run faster than the one on the ground.

This effect was checked in 1925, not for the earth, but for the white dwarf companion of the star Sirius (a white dwarf is an exceedingly dense star with a high gravitational field). Because time runs slower close to this white dwarf, radiation that is emitted from it will be changed in frequency. The change was calculated and observation agreed well with it.

In 1956 an effect called the Mossbauer effect was discovered that allowed scientists to verify this result here on earth. They were able to compare an atomic clock on the ground with one about 50 feet above it, and the difference in their rates was again in good agreement with Einstein's theory.

Just as the deflection of a light beam grazing the sun can be predicted using a simple "thought" experiment, so too can this slowing of time. According to the equivalence principle, an accelerated elevator generates a gravitational field. We can therefore consider two elevators in space accelerating at different rates. Assume that each has a clock in it. Assume further that the observers in the elevators communicate with one another using light signals: as each second is ticked off in the elevator that is accelerating the fastest, a pulse of light is sent to the other elevator. It is easy to see that because of the additional acceleration, the second elevator will receive the light pulses at intervals greater than 1 second, and therefore believe that the first elevator's clock is running slow.

If you have been observant, though, you will have detected a flaw in this thought experiment. As we just saw, the gravitational field decreases as we move upward above the earth; it is therefore slightly different over the height of an elevator (assuming the elevator is sitting on the earth). The gravitational field that is generated when you accelerate it upward, however, is uniform. This means that these two cases can be compared accurately only when the elevator is extremely small (the comparison is only exact when it is infinitely small).

In recent years new accurate tests of the theory have been made. In a 1971 test, Richard Keating of the U.S. Naval Observatory in Washington and Joseph Hafele of Washington University in St. Louis rode a commercial airliner around the earth, first in an east–west direction, then in a west–east direction. There are two effects on time in this case: a slowing due to the speed of the jet and a speeding up due to the decreased gravity. The clocks on the eastward flight lost, on the average, 59 nanoseconds (billionths of a second) compared to a predicted 49 nanoseconds, and those in the westward flight gained 273 nanoseconds compared to a prediction of 275 nanoseconds. In a similar experiment in 1976 an aircraft flew in circles around Chesapeake Bay. Again Einstein's theory was verified within experimental error.

As a result of these and other tests, Einstein's theory is now pretty well established. In the solar system or when gravity is weak, it gives results similar to Newton's theory, but when gravity is intense, as it is in the vicinity of a black hole (an object whose gravity is so strong no light can escape from it), Newton's theory breaks down and we must use general relativity. It must also be used when we deal with the structure of the overall universe.

Einstein became very famous as a result of his theory; he is probably the best known scientist who ever lived—one of the few most people in the street could readily identify from a photo. Despite the honors showered on him he remained unassuming and humble until his death in 1955. The publicity confused him but he took it in stride, goodnaturedly posing for photographers, artists, and sculptors. He posed so often, in fact, that when asked by a stranger who did not recognize him, what he did for a living, he replied, "I'm an artist's model."

In setting up his theory Einstein struggled with complex mathematical equations. He was able to make the breakthrough mostly, no doubt, because of his ability to concentrate so intensely on a problem. And the problems he was struggling with were not of the type encountered in the everyday world, or even in advanced university courses. Carl Seelig writes an interesting anecdote in his biography of Einstein showing the misconception some people have of his work.

Professor Einstein was sitting next to an 18-year-old girl at a party in America. When the conversation flagged, his neighbor asked: "What are you actually by profession?" "I devote myself to the study of physics," replied Einstein, whose hair was already white. "You mean to say you study physics at your age?" said the girl quite surprised. "I finished mine over a year ago."

The nature of Einstein's mind, his curiosity about the workings of nature, and his relentless need to probe them are evident in many of his statements. For example: "When I have no special problem to occupy my mind, I love to reconstruct proofs of mathematical and physical theories that have long been known to me. There is no special goal in this, merely an opportunity to indulge in the pleasant occupation of thinking."

Yet, after missing a question about the speed of sound on an exam he took in coming to America, he replied, "I don't clutter my mind with numbers I can easily look up in a book."

On another occasion he stated, "I sit here endlessly and think and calculate, hoping to unearth deep secrets."

As to what drove him to struggle so hard and so long, he replied, "It is really a puzzle what draws one to take one's work so devilishly seriously. For whom? For oneself?—one soon leaves, after all. For posterity? No, it remains a puzzle."

EXTREME WARPING

It might seem strange but Einstein never found the solution to his own equations. He showed that they went to Newton's equations in the appropriate limit, and he made several predictions as a result of them, but he never solved them. The first solution came, however, only a few months after the theory was published.

World War I was in progress during the winter of 1915–1916 when the astronomer Karl Schwarzschild read about the theory. He had never read about the theory at a university or in the quiet of his home as you might expect, however. Despite his academic credentials and age (in his 40s), he volunteered for war and had been sent to the Russian front. Within months he came down with a rare disease, and by the time he received Einstein's paper he was on his deathbed. Despite his weakened condition, he devised a solution for a spherical mass distribution and sent it to Einstein.

Einstein was pleased and surprised that a solution was obtained so soon after he had published the theory. He presented the solution at the next meeting of the Prussian Academy. Despite Einstein's joy at receiving a solution, he was perturbed at the nature of the solution. It told him that if the mass was sufficiently concentrated, something strange happened: the curvature or warping of space became so severe that the region inside a surface of a particular radius became cut off from the rest of the universe. Einstein did not like this aspect of the solution and worked off and on over the years to show that it had no physical counterpart in the real world.

It was while examining the details of this extreme curvature many years later that Einstein discovered something even stranger: as you approach the dense matter, space curves more and more, like a bottleneck or throat. But this throat did not end at the matter; according to the equations there was a mirror image throat attached to the other end of it. In effect, it was like a tunnel in space that got narrower and narrower, then finally began to open again. He wondered where this tunnel in space, or "space-time bridge" as he called it, led. He arrived at the conclusion that it could only lead to "another universe," whatever that meant. Einstein was not pleased with this result because it seemed possible that somebody could enter the tunnel at one end, pass through it, and emerge into this other universe. He was relieved, however, when further calculations revealed that a velocity greater than that of light was needed to pass through this tunnel, and according to special relativity this was impossible.

CHAPTER 3

Early Unified Field Theories

Just as Einstein became dissatisfied with special relativity short-ly after 1905, realizing it had to be generalized to all types of motion, so too did he become dissatisfied with general rela-tivity. There were several problems, all of which involved his field equation. Let us begin, then, by considering this equation. I do not want to scare you by writing it out in detail, so I will consider a simplified version that was stressed by one of my professors in first-year university. Looking down at the anxious faces before him, mine included, on the first day of lectures, he assured us that mathematics was really very simple. "All equa-tions are, after all, nothing more than $A = B$," he said. I'm not sure it was reassuring to most of the students, but it will at least help us now since it applies to Einstein's equation, although A and B are a little, or maybe I should say a lot, more complicated than most equations. They are both the mathematical quantities we talked about earlier called tensors.

From a simple point of view, then, Einstein's equation can be written: tensor A = tensor B, where tensor A describes the curvature of space and tensor B describes the matter that causes the curvature. In practice, B can also contain terms describing an electromagnetic field since electromagnetic fields represent en-ergy and energy is just another form of mass.

Einstein's dissatisfaction centered on tensor B. In his auto-biography he wrote, "The right side [of my equation] is a formal condensation of all the things whose comprehension in the sense of a field theory is still problematic. Not for a moment did I doubt this formulation was surely [temporary]. . . ." He introduced the tensor so that the equation would be complete and calculations could be made, but felt strongly that this was not its final form. "The left side is a cathedral made of marble, the right side is a house of wood and paper," he said many times.

The problem with the right side was that it was not a field term; it was a term that described matter. This meant that the overall equation was not a pure field equation. It was adulterated with matter, and this was distasteful to Einstein who felt the equation should be a pure field equation.

But the problem went much deeper than this: there were two fields known at the time and only one of them was contained in this equation. Besides the gravitational field, there was the electromagnetic field. It was in many respects similar to the gravitational field, yet distinctly different, and it was described by a completely different set of field equations known as Maxwell's equations. Even though the electromagnetic field entered the equations of general relativity on the right side as a source, it was not on the same footing as the gravitational field.

Why should there be two separate sets of equations for these two fields? Was it possible that the electromagnetic and gravitational fields are just different manifestations of the same field as electric and magnetic fields are? Are they connected in some way? If so, we should be able to write equations that describe both fields in the same manner. Einstein hoped to unify the two fields by incorporating the electromagnetic field into his equations of general relativity.

FARADAY, MAXWELL, AND THE
ELECTROMAGNETIC FIELD

To understand the difficulties we must take a closer look at the electromagnetic field. Even though it is all around us, associated with such things as TVs and radios, most people are less familiar with it than they are with the gravitational field.

Let us begin with electricity and magnetism. We know that a positive or negative charge is surrounded by a region of electric force (this force acts only on other charges) just as the earth is surrounded by a region of gravitational force. In both cases the strength of the force becomes weaker as you move away from the source. Magnetic fields are similar. You may remember performing an experiment in school in which you held a piece of paper over a magnet and placed iron filings on it. These filings oriented themselves in a way that showed the lines of force associated with the two poles of the magnet. We think of these lines as coming out of the north pole and entering the south.

Early scientists were familiar with electricity and magnetism, but thought of them as separate and distinct. That they were connected was shown by the Danish scientist H. C. Oersted. He was demonstrating electric currents to a class one day when a compass happened to be near one of the wires he had attached to the battery. He noticed that each time he attached the wire the compass needle moved. This could only happen if a magnetic field was being generated. Thus, by a simple fortunate accident, Oersted had demonstrated that an electric current creates a magnetic field.

It was then easy to go one step further: since a current is nothing more than charges in motion, and charges have electric fields associated with them, this meant that a changing electric field created a magnetic field. The news spread quickly; within months, scientists throughout Europe knew of Oersted's experiment and there was a frenzy of experimentation.

It might seem strange that someone could make such a significant discovery so easily, but of course scientific research was quite different in that day than it is now. I am sure many scientists today, as they hoist pieces of their apparatus together using huge cranes, sometimes long for the day when equipment was much smaller and brilliant discoveries could be more easily made.

Soon after Oersted's discovery, scientists began to wonder: If a changing electric field can create a magnetic field, is it possible that a changing magnetic field can create an electric field? And indeed, several years later this was shown to be the case. The critical experiment was performed by Michael Faraday.

Faraday, born in 1791 just outside London, moved shortly thereafter into the city. He grew up in poverty, spending much of his time playing in the streets. His education was meager, consisting of little more than the rudiments of reading and writing. When he was about 12 he was taken on as an errand boy by a bookbinder who a few years later made him his apprentice. This was the beginning of a gradual change in his life. The books that came in to be bound began to fascinate him; he wanted to read them all, but his poor education hampered him. He persisted, however, and his reading ability gradually improved. Some of the books were on science: electricity, magnetism, chemistry. Faraday copied as much from them in his spare time

Michael Faraday (1791–1867). (Courtesy AIP Niels Bohr Library.)

as he could before they were bound and shipped out. Several experiments were described in the books, and he spent as much of his small earnings as he could on apparatus to perform them.

The turning point in his life was his attendance of four public lectures given by Humphrey Davy, a well-known scientist of the day. Faraday was so enthralled he decided then and there that he wanted to be a scientist. He wrote up the notes he had taken during Davy's lectures, bound them as a book, and presented them to Davy, asking him for a job as an assistant. Davy was reluctant at first, assuring him that it was best to stay with the security of bookbinding. But shortly thereafter a position became vacant and Davy offered it to Faraday, who eagerly accepted it.

About a year later Faraday accompanied Davy on a lecture tour of Europe. He assisted him in setting up experiments, and for the first time had access to Davy's scientific library. He read voraciously, greatly expanding his already steadily increasing bank of knowledge. Soon after they were back in England, Davy suggested that Faraday begin experimenting on his own. This would mean, though, that he would have to strike out on his own, and at first Faraday was frightened. But meeting the challenge, he was soon hard at work, his ingenuity and persistence eventually surpassing those of Davy.

Like many before him, he began considering the possibility that changing magnetic lines of force could create electric lines of force. His classic experiment consisted of a loop of wire, an instrument to measure current, and a magnet. He noticed in moving the magnet through the wire that a current, and therefore an electric field of force, was produced.

This meant that not only did a changing magnetic field give rise to an electric field, but a changing electric field produced a magnetic field. Faraday went further than just demonstrating the effect; he wrote down a simple mathematical expression for the strength of the field generated. But he was not adept enough at mathematics to develop it.

One of Faraday's greatest contributions, strangely, was not

taken seriously for many years. Like many others working in the field, Faraday was genuinely puzzled by the nature of the electric and magnetic lines of force. What exactly were they? Mathematicians considered this force to be similar to the gravitational field—a kind of action-at-a-distance force. But Faraday was not satisfied with this; he introduced the idea of a "field." Lines could be drawn to represent this field; the closer the lines were together, the stronger the field. And these lines were not merely a geometrical structure for visualizing the field; as far as Faraday was concerned, the field had a physical reality. The mathematicians did not accept his view, but, of course, he did not accept theirs either.

Faraday continued to experiment and lecture as he grew older, but his health declined steadily. In 1841 he became so weak from overwork he had to take 4 years off. His fame, however, had already spread throughout Europe. Elected to the Royal Society, he was asked to become its president, but declined, saying, "I must remain plain Michael Faraday to the end." He came back to work in 1845 and continued to perform important experiments, but gradually weakened until his death in 1867.

Faraday was not a mathematician, and was not able to put his discoveries in mathematical form. But eventually his work came to the attention of James Clerk Maxwell, the greatest mathematical physicist of that day. Maxwell was 40 years younger than Faraday—born in the same year that Faraday announced the results of his famous experiment on the induction of an electric current by a moving magnetic field.

Maxwell's youth was quite different from Faraday's. His parents were fairly wealthy and he was brought up at Glenlair, an estate not far from Edinburgh. Like Newton and Galileo before him, he was fascinated by mechanical things, and as a youth built many ingenious mechanical toys. He remained at Glenlair until he was 10 when he went to the Edinburgh Academy to begin his education. His mathematical abilities soon began to show; at 14 he won the Academy's mathematical medal

James Clerk Maxwell (1831–1879). (Courtesy AIP Niels Bohr Library.)

for a paper describing a technique for drawing oval curves. The paper was read before the prestigious Edinburgh Royal Society.

Ten years later he entered the University of Edinburgh. That he spent only 6 years in school prior to going to university never hampered him in any way. By the time he was 16 he was, like Einstein, already thinking about mathematical problems well beyond his years. He read ravenously, and had an ability for intense concentration, being teased occasionally about living in a "different world." Sometimes at the dinner table he would drift off, immune to the conversation around him, as he performed a simple experiment in the projection of light or sound with some of the eating utensils.

While at the University of Edinburgh he had two more papers read before the Royal Society—a significant achievement for someone so young. In 1850 he went to Cambridge University, and was soon preparing for the highly competitive mathematical examinations—the tripos. Despite a severe illness just before the exams he took them with his legs and feet wrapped in blankets to ward off the cold, but still finished second.

Maxwell was described by his colleagues at Cambridge as congenial and brilliant—but different. He was always experimenting on something or trying to perform things in a different way. This even extended to his sleep. He became convinced it was best to split up his sleep, and for a while he slept from 5 to 9 every evening, then studied diligently from 10 to 2. He then exercised for half an hour by running up and down the flights of stairs in his dormitory. Needless to say his colleagues were not happy about this. Finally he slept again from 2:30 until 7.

After graduation Maxwell stayed on at Cambridge, lecturing and performing experiments. It was during this time that he was introduced to Faraday's work. Having heard of the controversy between Faraday's idea of a "field" and the mathematician's "action-at-a-distance," Maxwell approached the work cautiously. He wrote, ". . . before I began the study of electricity I resolved to read no mathematics on the subject till I had read through Faraday's *Experimental Researches in Electricity*. I was aware . . . that neither he nor they were satisfied with each other's language."

Maxwell soon became fascinated with Faraday's idea of a field, writing later in his book *A Treatise on Electricity and Magnetism:* "Faraday, in his mind's eye, saw lines of force traversing all space where the mathematicians saw centers of force attracting at a distance: Faraday sought the seat of the phenomenon in real action going on in the medium, they were satisfied that they had found it in a power of action-at-a-distance imprisoned on the electric fluid."

Maxwell decided to develop Faraday's ideas. He started by examining the four major known facts of electricity and magnetism:

1. Electric charges attract or repel and the force is of the same type (inverse square) as the law of gravity.
2. A moving charge or current produces a magnetic field (Oersted's law). Or since a charge has a magnetic field associated with it, we can say a moving electric field produces a magnetic field.
3. A moving magnet produces a current, and therefore an electric field (Faraday's law).
4. An electric current in one circuit can produce an electric current in a nearby circuit.

He soon realized that not only was there a kind of beauty in the field concept but that it could readily be put into the language of mathematics. His first efforts centered around the analogy between the field lines and the idea of flow that was used in the science that dealt with fluid flow, namely, hydrodynamics. Using the methods of hydrodynamics he introduced "tubes of flow," similar to tubes that carry water, only in this case the tubes carried an electric fluid. The velocity of the fluid represented electrical force, the difference in pressure from point to point, electrical potential (voltage difference).

He sent his first paper, "On Faraday's Laws of Force," to Faraday, who was shocked at first by the complex mathematical methods that were being brought to bear on the problem, but after studying it, was pleased with the new insight they provided. Maxwell published a second paper, "On the Physical Lines of Force," but shortly thereafter was interrupted by another problem.

Cambridge University had offered a prize for anyone who could determine the physical properties of Saturn's rings. It was a problem that challenged Maxwell's imagination and he was soon deeply involved. Over a period of 2 years he made numerous calculations that led him to conclude that the rings could not be solid, liquid, or gaseous; to remain stable they had to be composed of small particles, each in its own individual orbit around Saturn. His essay easily won the prize, and we now know that his view is correct.

Soon after he was awarded the prize he was back again working on the problem of electric and magnetic fields. Finally he was able to write each of the basic properties of the fields in mathematical form. The result was four equations. He soon found, though, that there was an internal inconsistency between the equations; to get around it he had to add a term to one of them.

Maxwell analyzed this term and found that it corresponded to a new type of electrical current—what we now call a displacement current. When an electrical force is impressed on a particle of electricity in an insulator, it would not normally be torn loose, but would instead be slightly displaced. When the electrical force was released, the particle would return to its equilibrium position and oscillate briefly around it. This meant that if we continually varied the electrical force in a periodic way (i.e., on–off–on–off), we could produce an oscillating or alternating current that would, in effect, move through the insulator. It is hard today to overstate the importance of this discovery.

Maxwell's four equations are still the basis of everything electrical or magnetic in nature, and they rank as one of the most important contributions of all time to physics. Electricity and magnetism were now completely unified, the relationship between them becoming particularly clear within the framework of the equations.

But Maxwell was not satisfied with merely writing the equations. He began examining some of their interrelationships and found that by an appropriate combination he could obtain equations that described waves—waves of electricity and magnetism. An oscillating charge would give rise to an oscillating magnetic field, and so on. What was important, though, was the way this combined electric–magnetic field acted: it would leave the oscillating charge and move off into space thereby beginning an independent existence.

The next time you watch TV or listen to the radio you can thank Maxwell for his insights. For it was this insight that gave rise to them. A signal in the form of electrons is sent to an antenna above a TV or radio station. These electrons accelerate back and forth on the antenna generating an electromagnetic wave that moves off into space. It is this wave that is picked up by your TV or radio. A simple representation of it is shown in the figure below: an oscillating electric field in one plane with an oscillating magnetic field perpendicular to it. The direction of propagation of the wave is perpendicular to both fields.

Once Maxwell had postulated the existence of such waves, he had to find out how fast they traveled. Using an ingenious balance, he made some accurate measurements that allowed him to determine their velocity. The result: electromagnetic waves traveled at the same speed as light waves.

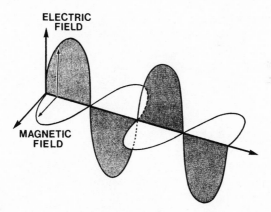

A simple representation of an electromagnetic field showing the electric and magnetic fields.

Was this a coincidence? Maxwell was sure it was not; if they both traveled at the same velocity, it was reasonable to assume they were closely related. Maxwell therefore took the bold step of postulating (correctly) that light was an electromagnetic wave and his set of equations also applied to it. Thus, we had another unification: electromagnetic phenomena and light.

It is now perh.ps easy to see why Einstein was so convinced that gravity and electromagnetism could be unified. If the unification of the electric and magnetic fields, then the unification of them with optical phenomena (light) are any indication it almost seems as if nature wants to unify. Although Maxwell showed that electromagnetic waves should exist, he did not live to see their discovery. Ten years after he died, Heinrich Hertz discovered the first nonoptical electromagnetic waves—radio waves. Today we know there is a whole spectrum of these waves, ranging from long-wavelength radio waves through microwaves and infrared to visible light. With even shorter wavelengths we have ultraviolet waves, X rays, and γ rays. The discovery of the laws that govern these waves was a monumental achievement. Most of our modern conveniences owe their existence to them.

In 1871 Maxwell accepted a chair at Cambridge University and for the next few years devoted himself to setting up what was to become the best equipped and most famous laboratory in Europe—the Cavendish Lab (named for Henry Cavendish, a famous scientist who worked earlier at Cambridge). He also spent several years editing Cavendish's unpublished papers. They were published in a two-volume work in 1879.

Everyone who knew Maxwell described him as friendly and completely unselfish—he once went without sleep or food for days as he nursed his ailing wife. As he grew older, however, he became more withdrawn and frequently seemed depressed. His friends tried to help but to no avail. The reason for his behavior later became known: he had cancer. For 2 years, he said nothing to anybody, and he did nothing about it. Finally the pain became unbearable and he was taken back to Glenlair where he died within 2 weeks. The greatest mathematical physicist of the period was dead, but in the same year—1879—another even greater one was born: Albert Einstein.

EARLY ATTEMPTS AT UNIFICATION

With Maxwell's discovery of the laws governing electromagnetic fields and Einstein's discovery of the field equations of gravitation, we had two important, but unrelated theories. Let us take a few minutes to compare these fields. There are similarities between them, but at the same time there are significant differences. Both require a source; the source of the gravitational field is matter and the source of the electromagnetic field is electrical charge. If we oscillate a charge, the changing electric field around it generates a magnetic field and the combination moves off into space as an electromagnetic wave. In the same way, if we oscillate matter a gravitational wave moves off into space. But in the case of the electromagnetic field there are two types of sources—positive and negative charges. There is no corresponding analogy in gravity; matter comes only in one form.

The two fields are similar, however, in the way they drop off in magnitude around their sources. The electric field gets larger and larger the closer you get to the source generating it. To see an important consequence of this, consider the electron. The closer you get to it, the higher is the electric field until right at its center, according to the theory, the field becomes infinite. We say there is a *singularity* at this position. A similar situation occurs in the case of gravity; there is a singularity at the center of the mass.

This was an aspect of the theory that Einstein thoroughly disliked. He did not believe in field singularities and felt we should somehow be able to get rid of them. "Material particles have no part in a field theory," he wrote in a *Scientific American* article in 1950. (This is, of course, related to the problem of "sources" we discussed earlier in relation to his field equation.)

The similarity of the electromagnetic and gravitational fields, and the possibility of unification was noticed even before Einstein gave it his attention. The first person to attempt a unification was the German physicist Hermann Weyl. Weyl considered an aspect of general relativity that we talked about earlier in relation to Riemann geometry—the nonpreservation of direction in a curved space. We can illustrate this best by considering the surface of the earth, which we know is a two-dimensional Riemann surface. If two airplanes, initially at some distance apart, take off from the equator and fly on a parallel course to the north pole, their flight paths will not remain parallel. In fact, we know they will cross at the pole, which means that although they started off flying in exactly the same direction, namely north, they will be flying in distinctly different directions when they reach the pole (and even before). You can easily verify this by looking at the lines of longitude on a globe. This means that direction is not preserved in curved space.

Weyl decided to consider the possibility that length was also not preserved. This would mean that not only would the direction of the airplane change as it moved through space, but its length would also change. To effect this change mathematically, Weyl had to make a slight modification in the structure of general relativity. He assumed that in addition to the usual metric (set of numbers or variables) that described the gravitational field, there was another one related to length. It might seem that this was a little like somebody trying to shoot a black cat in a dark alley. Lengths do not change in the real world, just because they move in different paths. But amazingly when the result was analyzed Maxwell's equations mysteriously appeared. It almost seemed as if a little bit of magic had occurred and scientists quickly became interested in the miracle.

Weyl had produced, by slightly modifying general relativity, a theory that described both the electromagnetic field and the gravitational field. Did this mean a unification had been accomplished? Many people thought so at first, but with detailed analysis the theory was shown to be flawed. Einstein was the first to put his finger on the flaw.

The problem centered around the concept of length. In relativity the only meaningful length is a length of space-time. In other words, a length always includes a time part, or time interval. This meant that if two different objects took two different routes to the same point, not only would their lengths change but also the related time interval would change. This time interval might correspond, for example, to the vibrational frequency of an atom. This would mean that if two identical atoms took different routes in space to the same point, they would vibrate at different frequencies. We know this does not happen; if it did we would not see sharp spectral lines from distant stars, and we do.

Weyl soon acknowledged the flaw and laid his theory to rest. It may have been a failure (actually it was not an entire failure; a similar idea is used today in modern field theory), but it did accomplish something important: it got people interested in the possibility that the electromagnetic and gravitational field could be unified. Einstein soon began working on an alternative theory, as did others.

In 1921 another interesting unification was attempted by Kaluza in Germany. Kaluza showed that if Einstein's equations were written in five dimensions instead of four, the same miracle happened: Maxwell's equations appeared within the theory. Einstein surely must have asked himself at this stage, "Is God playing tricks on us?" We now had two theories—both containing the equations of the electromagnetic field, namely Maxwell's equations, in addition to Einstein's gravitational field equations. There was considerable interest in Kaluza's theory for several years. But there was a major problem: the real world had only four dimensions—three of space and one of time. What was the significance of the fifth dimension? Where did it fit in? It became obvious to Kaluza that he would somehow have to suppress it when a correspondence was made with the real world. He did this with what we call a mathematical projection—like the two-dimensional shadow on a wall from a three-dimensional object. This brought it into accord with nature, but difficulties remained.

In 1926 the Swedish physicist Oscar Klein extended the theory. He suggested that the fifth dimension was not observable physically because it was in the form of a loop that was wound up so tightly we could not see it. Others, including Einstein, worked on the theory but gradually interest in it waned. One of the chief difficulties was that it did not predict anything new. It gave Einstein's equations along with Maxwell's but it seemed to give little else. Recently, however, there has been a resurgence of interest in the theory and a number of scientists believe it may eventually lead to an important breakthrough. Abdus Salam, a prominent theoretical physicist, referred to it recently as "one of the four major developments in the realization of Einstein's dream." A modern version of the theory that has 11 dimensions and is connected with another important theory called supergravity has generated considerable interest lately. It will be discussed later.

Another unified field theory—geometrodynamics—was put forward in 1957 by Misner and Wheeler. Actually, the same idea had been used by Rainach in 1925 but Misner and Wheeler were unaware of his work. It is sometimes referred to as the "already unified field theory." To see the significance of the name, consider Einstein's equations, which I earlier abbreviated as $A = B$. Wheeler and Misner found that with a little juggling they could put B in the same form as A. With this modification the electromagnetic field, which was previously inside B, becomes a geometric term like A. The new form of the equation contained both electromagnetic and gravitational fields and had no sources.

It might seem that an equation with no sources would present a problem. Where would the two fields come from? Misner and Wheeler got around this ingeniously by using an idea that Einstein had introduced earlier: wormholes in space (Einstein called them bridges in space). Consider the wormholes shown in the figure on the next page. Field lines flow through them,

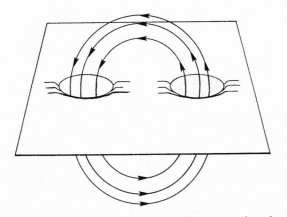

A simple representation of a wormhole in space. The field lines enter through one hole and exit through the other.

entering at one end and exiting at the other. Looking down on these holes from the surface they would appear as sources: one a positive source (lines coming out) and the other a negative source (lines going in). But the source of the electromagnetic and gravitational fields are particles, thus in introducing these tubes we would be replacing particles with wormholes. There would be no such thing as matter, only wormholes in space. This is, of course, exactly what Einstein wanted. He abhorred singularities, and the usual sources—particles—created singularities. Wormholes got around this problem.

For several years many people worked on unified field theories based on generalizations of general relativity. But gradually, as the difficulties compounded and hope faded, one after the other left the field for greener pastures. W. Pauli, who worked on several theories, eventually became exasperated in his attempts. He was sure that such a theory was not possible. "What God hath put asunder, no man can form together," he once said in disgust.

EINSTEIN'S UNIFIED FIELD THEORY

In 1925 Einstein began working on a theory that he would continue working on, with occasional diversion, throughout his life. The main problem that concerned Einstein, namely the nature of the sources, was actually around before Einstein directed his attention to it. How, for example, did particles hold themselves together? The electron, for example, is made up of negative electrical charge. But negative charge repels nearby negative charge, so it seems the electron should, in a sense, repel itself. One section of the charge should repel a neighboring section—the electron should explode!

In a sense this is still a problem, even today. We have not found a satisfactory theory for the interior forces of the electron, but now get around the difficulty by assuming the electron has no interior. It is a point charge with no dimensions, and thus cannot repel itself. Experiment seems to bear this out.

Like Weyl and Kaluza, Einstein felt that a unified field theory had to come from a generalization of general relativity. Weyl had generalized by adding a metric, Kaluza by adding a dimension. Einstein looked for other possibilities, and he found one—a particularly satisfying and natural one as far as he was concerned. General relativity was a symmetric theory; in other words, the metric was symmetric, much in the same way your body is symmetric about a vertical line down its center. You have two eyes, two ears, and so on. Einstein decided to see what would happen if he did away with this symmetry, creating in its place a nonsymmetric theory. Once again, Maxwell's equations miraculously appeared, and once again it seemed as if a truly unified field theory was at hand. Einstein worked on the theory for a while, but abandoned it briefly a few years later to work on another theory—a variation on Weyl's theory.

In early 1929 Einstein became convinced this variation was the correct theory. The news leaked to the press and soon excited headlines appeared in papers around the world. Einstein's new theory was heralded as a major breakthrough in science.

But Einstein knew it had not yet been tested, and that would likely take several years. The publicity was quite embarrassing to him, particularly when it turned out he discovered the theory was lacking.

One of the major difficulties of Einstein's attempts at unification was now beginning to come to the forefront. A few years earlier quantum theory had been discovered, and the results it gave seemed to be in good agreement with experiment. If Einstein's unified field theory was to be the ultimate unified theory of the universe, it would somehow have to incorporate quantum theory.

Einstein was not happy with quantum mechanics and its statistical approach to the problems of the microworld. He was sure his generalization would somehow overcome the ideas of probability and uncertainty. But quantum mechanics developed quickly, and within a few years most physical phenomena (in the microworld) were under its wing, explained in terms of its strange new language. Scientists began to abandon general relativity, most climbing aboard the new and more promising bandwagon of quantum mechanics. Einstein was soon virtually alone in his quest for a generalization of his theory.

In 1932 an offer came from the Princeton Institute of Advanced Study in the United States, and a year later Einstein crossed the ocean for the last time (he had been in the United States briefly earlier). But he was no longer at the forefront of physics and the attitude of many of his colleagues began to grieve him to some extent: they now began to look upon him as something of an antique caught in a classical rut. His disdain for the now-popular quantum theory puzzled them. But he remained undaunted, enduring with good humor their silent jeers. He knew how he appeared to them, saying in 1954, "I must seem like an ostrich who forever buries his head in the relativistic sand in order not to face the evil quanta."

Einstein worked with various collaborators while at Princeton: Hoffman, Infeld, Strauss, and Bargmann. And many times he thought he had reached the ultimate goal, only to find a few days or weeks later that his new house of mathematical equations had collapsed. Strauss wrote, "We continued working on [a particular theory] for nine months. Then one evening I found a class of solutions which in the light of the next morning showed the theory could not have physical significance." Strauss was almost heartbroken over the setback, but related, ". . . the following morning Einstein had forgotten about our setback and was already thinking about a new theory."

New ideas still came to Einstein but not in the way they did in his youth. Furthermore, the difficulties with the new theory were even greater than they were when he was formulating general relativity. He had a number of guidelines in the case of the earlier theory, but now he was groping in a mathematical jungle of complex equations with little light or guidance. It was more a hit-and-miss proposition: try it, if it did not work, try something else.

Someone once asked him if the tremendous effort he had put into the search had served any useful purpose. "At least I know 99 ways that don't work," he replied. But he always felt compelled to search on. "I agree the chance of success is small," he said, "but the attempt must be made . . . it is my duty."

His search while at Princeton centered on his earlier non-symmetric theory. He had devised two sets of equations—each a possibility. But again each developed difficulties. Leopold Infeld showed that particles described by one set would not interact properly—they did not satisfy the ordinary well-known laws of electricity and magnetism. Later Callaway showed that the same thing applied to the other set.

Einstein was convinced, however, that these equations were only a first step. Somehow they could be changed slightly, or modified. There had to be a way. As he pushed on in his search, grief struck many times as he lost close friends. Three years after he came to Princeton, both his wife and his friend Grossman died. Then in 1946 his sister Maja, who had been extremely close to him, suffered a stroke; she lingered on until 1951.

Einstein a few years before his death. (Courtesy AIP Niels Bohr Library, Dorothy Davis Locanthi Collection.)

By the mid-1950s, even Einstein began to have doubts. He had tried so hard and yet had not achieved his goal. Not long before his death he admitted reluctantly, "It appears dubious whether a field theory can account for the atomistic structure of matter and radiation as well as quantum phenomena." But his final lack of conviction did not stop him from trying. Even while he was on his deathbed he had his pencil and pad in hand.

On April 13, 1955, he was taken to the Princeton hospital with severe cramps. He knew it was the end but he asked for his glasses and notes and continued working. Friends visited him in the hospital. Looking at their downcast, drawn faces he said, "Don't look so sad . . . everybody has to die." He passed away on April 18, his dream unfulfilled.

Many of the problems of Einstein's new theory centered around the mathematics rather than the physical interpretation. The mathematical difficulties were so great that we might ask if perhaps a new mathematical technique is what was really needed. If we look back in history at great scientific advances, we see that they are frequently associated with the discovery of new mathematical techniques. Newton, for example, made most of his advances after discovering calculus. And Einstein would not have been able to formulate his general theory of relativity without tensor analysis, which had been formulated only a few years earlier.

Is a new mathematical technique needed in order to overcome the difficulties? Since we still have not overcome them, we do not know. One of the things that was a difficulty in Einstein's theory has, however, been overcome: the difficulty with the strange way particles interacted. They did not interact in accord with basic laws. Einstein himself had tried the method that was used, but had rejected it. He felt that the equations, like those of general relativity, should be simple—they should have a certain scientific beauty about them. He therefore abhorred the thought of adding terms to them. But in 1952 B. Kursunoglu added a term and arrived at a theory that overcame this difficulty, and in 1954 a similar, but different, theory was formulated by W. B. Bonner. Solutions were obtained for these theories but neither is generally considered to be successful.

There are still basic difficulties. First of all, if such theories are to explain the various particles, they must explain why they have different properties (e.g., charge, mass)—and they do not. Other theories, related to quantum theory, which we will talk about later, come fairly close to doing this. But there is another serious difficulty that we have so far overlooked. We have talked only about the gravitational and electromagnetic fields.

These were the only two fields known most of the time while Einstein was working on the theory, but as we saw in Chapter 1, we now know there are two other fields—the strong and weak nuclear. For the theory to be truly unified, these other fields would have to be included.

CHAPTER 4

Star Death

As we have seen, scientists tried to extend general relativity in an effort to forge a unification of the gravitational and electromagnetic fields. The attempts were ingenious and, in some cases, tantalizingly close to the goal, but the unification of these two fields has not yet been achieved.

Suppose, though, that we did achieve such a unification. Would our problems be over? Would we have a unified field theory? Hardly. There are two fields that are not within the theory; moreover, we want a theory that never breaks down, and general relativity does. We know, for example, that it does not work in the realm of atoms, but aside from that there is another condition under which it does not work: at exceedingly high densities. And I am not talking about the moderately high densities we encounter in everyday life. Far from it. These densities occur only under the most extreme cosmological conditions.

Our task, then, is to see where and understand why this breakdown occurs, and the best place to start is with stars—plain, ordinary stars. Our sun is, of course, a star, one we are particularly familiar with so we will begin with it. Through a proper filter it looks like a smooth, shiny disk, a place of tranquility, but if you examine its surface more closely with an appropriate instrument you see that it is actually a sea of hot seething gas in constant agitation. Occasionally the turmoil becomes so intense that gigantic arching plumes stream upward at tremendous velocities. Making their way along magnetic field lines, they reach out for thousands of miles, then rain back on the surface.

But not all the material that is thrown out rains back. Intense flares occasionally burst forth spewing particles into space that leave the vicinity of the sun and stream throughout the solar system.

What causes this tremendous turmoil? For an answer we must look beneath the surface to the interior of the sun. Here we find layers of gas that are even hotter than those of the surface. As you move closer to the center, in fact, they get increasingly hotter. The Swiss astronomer Jacob Emden was the first to suggest that our sun might consist of hot gas throughout, but it was the British astrophysicist Arthur Eddington who developed the idea and gave us our first mathematical model of the interior of a star.

EDDINGTON

Eddington was an enigma. He was a genius and one of the giants of astronomy, yet in his later years he did strange things. Like many great scientists after him, he eventually turned his attention to the unification of general relativity and quantum theory. But the culmination of his efforts, a book titled *Fundamental Theory*, was understood by few, if any, and is today looked upon with little more than curiosity.

Eddington, born in 1882 in Westmorland, England, was the son of Quakers. His father, the headmaster at a local school, died when the boy was only 2, and he was raised by his mother. His mathematical abilities were evident at an early age: he learned the multiplication tables up to 24 × 24 before he could read, and by the time he was 10 he was glued to a telescope night after night, fascinated by what he saw in the sky.

He won honors in high school, then capped them off with a scholarship to Manchester University. To his surprise, though, when he arrived at Manchester he found he was too young (15) to enter. Fortunately, someone had the foresight to relax the rules in his case. From Manchester he went to Trinity College at Cambridge, famous for its mathematical tripos (the same exams Maxwell took earlier). Eddington took them at the end of his second year, ranked first, and became "senior wrangler"—the only person to accomplish the feat in such a short time.

Upon graduation he worked for a while in experimental physics at Cavendish Laboratory but soon became bored. It was theoretical physics he was interested in—not toying with laboratory apparatus that never seemed to work. So when a position became available at Greenwich Observatory in 1906 he gladly accepted it.

Despite his ineptness in the lab, he did become an excellent observer and was soon engaged in serious astronomical research. The Dutch astronomer Jacobus Kapteyn had just organized a worldwide effort to study our galaxy. Some of his first results indicated that there were two distinct streams of stars in the vicinity of the sun. Eddington helped show that this was due to the rotation of our galaxy. Astronomers were still uncertain, though, not only of the detailed structure of our galaxy, but of whether there were other galaxies in the universe. Many "fuzzy objects" could be seen in the telescope, some of them elliptical and a few quite irregular, but what were they: "island universes" of stars like our galaxy or just gaseous blobs? Eddington favored the idea that they were island universes, and later this was shown to be true of most of them.

In 1912 the Plumian Chair at Cambridge University became vacant upon the death of G. H. Darwin, son of Charles Darwin. Everyone was sure it would go to James Jeans, one of Darwin's students, but the committee chose Eddington. This was particularly surprising in that Eddington was only 30 at the time, 5 years younger than Jeans. Jeans, naturally, was disappointed and for many years considered Eddington a serious rival in almost everything he did. Colorful and sometimes quite heated public debates between the two men took place over the next few decades.

In the same year that Eddington became Plumian professor at Cambridge, an important diagram called the Hertzsprung–

Russell diagram in honor of its discoverers was published. It was a plot of a star's true or absolute brightness against its surface temperature. Most of the stars in such a plot tend to lie along a diagonal, indicating an approximate relationship between the two variables. Eddington became convinced that this diagram was the key that would open the door to an understanding of the star's interior, and, indeed, it turned out to be a Rosetta stone.

At that time, though, almost nothing was known about the interior of a star. Emden had suggested they were gaseous throughout but many astronomers were convinced that they consisted of some sort of incompressible fluid—perhaps like hot glue. But no one had any idea how hot they were deep inside. In fact, it was a question that had not even crossed most astronomer's minds. Actually, it was not stellar interiors that first interested Eddington; his involvement with interiors developed indirectly. He set out to explain the strange pulsations of Cepheid variables—large stars that undergo periodic changes in brightness—but soon found that almost nothing was known about their internal structure, or that of any type of star, for that matter. He therefore had to consider this problem first.

Eddington is on the extreme right. H. N. Russell (of HR diagram fame) is third from the left. (Courtesy AIP Niels Bohr Library, Margaret Russell Edmondson Collection.)

Starting with the idea that they were gaseous throughout, Eddington decided to see what conditions would be required for stable equilibrium. Gravity would obviously create a tremendous inward pull that would somehow have to be countered by an outward force. This force, it seemed, would come from the gas pressure. Eddington's stroke of genius, though, was in realizing that in addition to ordinary gas pressure there would be radiation pressure. It was well known, for example, that ordinary light exerted a pressure, and inside a star where the radiation levels were exceedingly high this pressure would be substantial. Eddington found, in fact, that it was this pressure, and not the gas pressure, that was primarily responsible for holding the star up (keeping it extended). He calculated several properties of the star based on this assumption and found that his results agreed well with observation.

One of his most important calculations was the determination of the core temperature of a star. He was amazed at how high it was—15 million degrees. At this temperature the atoms had to be ionized (electrons separated from nuclei). Eddington then went on to do the impossible: he created a complete mathematical model of the interior of a star. He must have chuckled to himself a little as he remembered a statement that had been made only a few years earlier by a famous astronomer: "We will never know what goes on inside a star. It is forever beyond us." In 1926 he published his results in the now-classic book, *The Internal Constitution of the Stars*.

Eddington remained a bachelor throughout his life; he kept in shape by golfing, hiking, and bicycle riding. Bicycle riding was his passion: he occasionally rode over 100 miles a day, and even when he was 60 he still rode up to 80 miles in a day. He relaxed by reading detective stories and working crossword puzzles. He was shy and generally uncomfortable in the presence of women and seemed to have little interest in them physically; aside from his mother and sister who kept house for him, his relationships with the opposite sex were strictly casual.

As an academic lecturer, Eddington was dull; he would walk into a classroom, pull his tattered, now-famous, text from a huge pocket inside his jacket, and begin, and soon most of his students would be nodding. Yet strangely his popular talks were a huge success. He spent considerable time preparing them, and people flocked to listen to him. They were, in fact, perhaps too organized, for when he was occasionally forced to deviate from them he sometimes became flustered. Questions bothered him; he would become nervous and tongue-tied in answering them. But all in all both his talks and his popular books were a tremendous success, and he seemed to enjoy popularizing science. He is probably best known for this aspect of his life.

After Eddington determined the internal structure of stars, he spent a considerable amount of time wondering about the source of their energy. It was obvious that they were giving off a tremendous amount of energy and had been doing so for millions of years. After his colleague F. W. Aston showed that four hydrogen atoms weighed more than one helium atom (stars are made up mostly of hydrogen and helium, and helium atoms are, in effect, made up of four hydrogen atoms), he began considering the possibility of the conversion of mass to energy. Was it possible that some of the mass of the star was being converted to energy? If so, a tremendous amount of energy would be released according to an equation Einstein had published several years earlier. Eddington became convinced it was this conversion that powered the stars.

Eddington had many honors showered on him throughout his life: 12 honorary degrees, the gold medal of the Royal Astronomical Society, he was knighted in 1930, and in 1938 he received the prestigious Order of Merit.

His death was sudden and unexpected. Late in 1944 his health began to deteriorate; he tried to continue bicycling but soon had to give it up. He kept the pain to himself for a while but eventually became so weak he was forced to see a doctor. The doctor operated immediately and found an incurable cancer. He died shortly thereafter.

Eddington will be remembered for his tremendous insight

and the great strides he made in stellar astronomy. Yet, as we will see later, he was also a person who held up progress in astrophysics by stubbornly challenging new and important ideas.

THE LIFE CYCLE OF A STAR

With Eddington's results, astronomers could tackle the problem of how a star evolved. We now know that there are, indeed, nuclear reactions going on in the core of a star, which power the star; astronomers refer to the region where they take place as the thermonuclear furnace. Hans Bethe worked out the details of these reactions in the case of the sun in 1938. He showed that hydrogen is converted into helium in a series of reactions that generate a tremendous amount of energy. Over a long period of time this energy makes its way to the surface and is radiated into space.

Our sun is, of course, just a star, like the hundreds you see in the sky on any clear night. It is a million times as big as the earth, yet it is only an average-sized star. There are red giant stars thousands of times as big and tiny white dwarfs barely larger than the earth. Through an extensive study of all these stars we have arrived at an acceptable theory of how our sun and the planets around it developed. It is assumed that all stars and their planets (if any) developed in this same way.

About 5 billion years ago, according to our best estimates, there was a gigantic cloud of gas made up mostly of hydrogen but with some helium and a small amount (approximately 1%) of heavier elements. This cloud, called the *solar nebula,* was cool, irregular in shape, and it spun slowly. As self-gravity pulled it inward, its spin increased and it took on a roughly spherical shape. Eventually the outward force created by the spin stopped the material in one plane from falling any farther. The gas in all other directions, however, continued to fall until the cloud became a giant disk with a bulge at the center.

As the gas of the bulge condensed, it began to heat near the center, and the resulting radiation began moving through the nebulous gas that surrounded it. The bulge, referred to as a protostar at this stage, would eventually become our sun. The radiation that was emitted heated the regions it passed through, but it had difficulty penetrating the dense clouds and a gradual dropoff in temperature resulted.

Tiny grains then began to condense out of the nebula. In the inner hotter regions grains of iron, nickel, and heavier elements formed; farther out where it was cooler silicon condensed, and farther still grains of methane, ammonia, and lighter elements condensed. These grains fell toward the central plane of the gaseous disk and eventually formed a gigantic ring inside of it, similar to Saturn's ring, only much larger. And like the particles in Saturn's rings the inner ones in this ring traveled faster than those farther out. This caused them to strike one another occasionally coalescing and gradually growing into small rocks. The rocks also interacted and coalesced until some of them resembled small asteroids, or planetesimals. The surfaces of these planetesimals were then pounded by the small debris that existed between them, and they in turn collided creating protoplanets—rocky cores surrounded by dense atmospheres of hydrogen and helium.

At this stage the system was beginning to resemble the solar system we know today but there was a major difference. It was still totally immersed in a gigantic cloud of hydrogen and helium. The temperature in the core of the protosun was, however, getting close to a critical value—15 million degrees. When it reached this magic number nuclear reactions were triggered, and the protosun became a star. But the sudden triggering had an overwhelming effect: a powerful explosive wave—a solar gale—rushed outward from the sun blowing the fog from the protoplanets and cleaning up the solar system. The inner planets were stripped down to their barren, rocky surfaces, but the outer ones, being larger and farther away, managed to retain most of their atmospheres. As we look at the solar system, today, we see that the larger outer planets still have immense hydrogen–helium atmospheres.

For millions of years the inner planets, earth included, were barren of atmospheres, just as the moon is today. But gradually their interiors heated as a result of radioactivity until they were molten. The molten lava and accompanying gas then began pushing upward, creating volcanoes. The gases expelled from the volcanoes began to accumulate and finally the earth had an atmosphere, but it was different from our present one, consisting mostly of methane, ammonia, nitrogen, and water vapor. Slowly, though, over the years it evolved into the one we have today. Then as water vapor condensed, oceans formed and eventually life appeared in them.

All stars develop in the same way. They begin as giant gas clouds, condensing as a result of gravity to protostars and when their core is sufficiently hot they generate nuclear reactions and become stars. Because most protostars, like our early sun, probably also had disks of matter around them, it seems reasonable that solar systems such as ours are common throughout the universe. And indeed we have strong evidence that some of our neighboring stars have dark objects circling them, even though we cannot see them directly.

Our sun is now in an equilibrium state. It is peacefully burning hydrogen, leaving helium as ash, and it will continue doing so for a few billion years. Eventually, though, it will run out of fuel, but even before this occurs dramatic things will happen. Astronomers know this because observation and computer models have allowed them to develop a detailed theory of stellar evolution. As a result they are now able to predict how stars will change over millions and even billions of years.

Because helium is heavier than hydrogen, it will accumulate at the center of our sun as hydrogen burns. The hydrogen will continue to burn in a shell around it, and as more hydrogen burns the sphere of helium will grow. But as the helium core gets larger the pressure at its center will increase, which will increase its temperature.

The increased temperature will make its presence felt first near the surface of the sun. It will begin to drive off the outer layers, and as they move outward they will cool. Our sun will begin to expand and slowly, over millions of years, temperatures on earth will increase. The polar caps will melt, raising the level of the oceans and thereby flooding most of the coastal cities. Temperatures in the equatorial regions will become unbearable and there will be an exodus to northern and southern latitudes. Then finally as these regions get too hot the exodus will continue to the poles.

With increasing temperature the oceans will begin to evaporate and cloudiness will increase. Finally the earth will be shrouded in a heavy fog with rain falling continuously. But the increased cloudiness will push temperatures even higher due to the greenhouse effect. This is the reason for Venus's high temperature: radiation from the sun passes through its cloud layer, but when it is reflected from the surface it is changed in wavelength and is therefore unable to get back out through the clouds. As it is reflected back and forth between the surface and the clouds, it generates a tremendous amount of heat.

Finally all life on our planet will be gone. The oceans will have dissipated into space and their surface will be barren. But temperatures will continue to increase and eventually the Earth's surface may become molten, for the sun's outer shell will begin to move outward like a gigantic solar storm. First it will consume Mercury, then Venus, then it will approach Earth, and an eerie red twilight will overcome us. But it will not devour the Earth; just before it reaches our orbit its outward expansion will cease.

Meanwhile, deep within the sun, at its core, helium is still piling up and the ever-increasing pressures are creating unheard-of temperatures. When the temperature reaches 100 million degrees, it will be hot enough to burn the helium. But the helium at this stage will be so rigidly packed that when it is ignited at the center it will act like a giant fuse. It will not be able to expand to compensate for the sudden reactions that are occurring and within seconds these reactions will race unchecked throughout the core. The entire core will be ripped apart in the ensuing explosion and the hydrogen-burning ring surrounding it will be struck with such a force that it will be blown entirely apart. The nuclear furnace will be extinguished.

It might seem that such an explosion, called the helium flash, would blow the sun apart, but it is so extended at this stage it will show no immediate evidence of the explosion at its surface. As time passes, though, the effects of the explosion will make themselves known. With the nuclear furnace out, no radiation will be able to filter outward, and eventually the outer shell of the sun will cool. It will continue to cool for millions of years as the helium falls slowly back to its original position. Finally, when it is all back in place it will begin burning peacefully at the center; the hydrogen surrounding it will also begin burning again in a shell. It will continue burning peacefully in this way for a few million years—a much shorter time than its original hydrogen-burning period.

But as the core continues to burn, its temperature will continue to increase and as a result the outermost layers will expand and cool even more. Eventually, they will be cool enough for electrons and nuclei to begin re-forming into atoms. Photons (particles of radiation) will be given off in this process, generating considerable heat. The process will soon get out of control and the sun's entire outer shell will be pushed off into space.

Looking around us now with a telescope we see many stars that are in the process of doing this. They are called planetary nebulae. The shell emitted by the sun will sweep by the earth and out past the gas giants, eventually dissipating off into space. With the loss of its outer cooler layers, the sun's surface temperature will skyrocket—going from a few thousand degrees to perhaps 50,000 degrees.

In its core the helium will continue to burn, and like hydrogen it will also leave ash—carbon and oxygen. Again the carbon and oxygen will be heavier than the helium and fall to the center. The helium will soon be burning in a shell around it. The

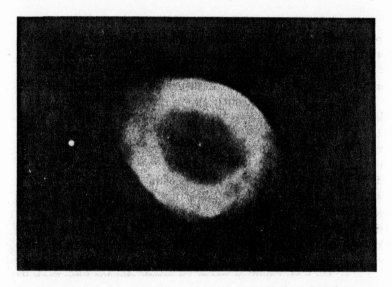

The Ring nebula in Lyra. A planetary nebula. (Courtesy National Optical Astronomy Observatories.)

carbon and oxygen will also burn if the temperature reaches 3 billion degrees, but the core of the sun will never get this high.

The sun will burn hydrogen and helium but it is not massive enough to burn anything else. What will eventually happen to it? Like all stars it will die, but its death will be slow and peaceful. Its nuclear furnace will go out and it will begin slowly contracting inward on itself. Over millions of years it will become increasingly dense, finally reaching densities of tons per cubic inch. It will become what is known as a white dwarf.

CHANDRASEKHAR AND THE WHITE DWARFS

The first white dwarf was discovered by Friedrich Bessel in the early 1840s. He noticed in studying the motion of Sirius over many years that its path through the sky was not a straight line as might be expected. It had a slight wobble that Bessel believed was caused by a nearby star—one he could not see. And in 1862 a tiny point of light was discovered next to Sirius by the American telescope maker Alvin Clark. We now refer to it as Sirius B; the brighter component is called Sirius A. Astronomers were puzzled by the tiny star, wondering if it was perhaps a dying star. They were astounded in 1915, however, when Walter Adams of Mt. Wilson Observatory was able to pass its light through a spectroscope and showed that its surface temperature was 8000°C (14,000°F).

How could such a tiny object have such a high temperature? With a surface that hot, it could not possibly be a dying star; also it had to be much smaller than originally believed. A simple calculation showed that its surface area was only about 1/2800 that of Sirius A; its density therefore had to be incredibly high—15 tons per cubic inch.

A strange object. But why was it so small and heavy? What was going on? The answers to these questions did not come until 1927 when Ralph Fowler of Cambridge University applied quantum theory to the problem. He realized that if the temperature was as high as the spectrum indicated, the atoms within the white dwarf would be completely dissociated; its interior would therefore consist of tiny nuclei floating in a sea of electrons. But electrons and nuclei, together as atoms, occupy a much greater volume than they do individually as particles. Strange as it may seem, atoms are mostly empty space.

"But my body is made up of atoms and this would imply that I'm mostly empty space," you might say. Moreover, if you can feel your hand, how can it be mostly empty space? True, it does seem as if your hand is relatively solid, but this is because the electrons whirling around the nucleus create a barrier, and that is what you feel. If you could examine the atoms closer, though, you would see that inside the electronic barrier there is mostly empty space—the nucleus only takes up about one-trillionth of the space here. This means that if the atoms dissoci-

ated into electrons and nuclei, and the pressures were sufficiently high, they would be compressed into a much smaller
volume. A star as large as our sun could be compressed into an
object not much larger than the earth.

But what finally holds the star's tremendous weight up?
There would have to be some sort of outward force holding back
the tremendous gravitational pull. According to a well-known
principle put forward by Wolfgang Pauli in 1925, each electron
takes up a certain volume, and cannot be compressed any smaller. When the white dwarf reaches the state where the electrons
are all compressed to their minimum volume, it can go no further. The mass of the star is then being held up by electron
pressure.

For years there were many unanswered questions about
white dwarfs. Did all stars become white dwarfs, and if not,
what happened to them? Subrahmanyan Chandrasekhar began
considering these problems shortly after the German physicist
Arnold Sommerfeld visited his undergraduate school in
Madras, India, in 1928. His interest in Fowler's work increased
and upon graduation he decided to go to Cambridge to work
under him. Like Fowler he applied quantum theory, but in addition he realized that at the high temperatures inside white
dwarfs the particles would have extremely high velocities and
special relativity would be needed.

Following Fowler he showed that electron pressure would
hold up a star with approximately the mass of the sun. It would
then remain stable for billions of years, slowly radiating its remaining energy to space, and in the process slowly cooling. But
when he looked at stars with greater mass than the sun he
discovered something strange: electron pressure could not hold
them up. There was a mass of about 1.4 solar masses, beyond
which the electrons were incapable of sustaining the load. We
refer to this as the critical mass.

When he arrived in England, Chandrasekhar discussed his
results with Fowler and another well-known astronomer, E. A.
Milne. Both were skeptical of the notion of a critical mass. What

S. Chandrasekhar (1910–). (Courtesy AIP Niels Bohr Library, Dorothy Davis Lo-canthi Collection.)

Chandrasekhar had, in effect, shown was that in large-mass stars the electron "gas" never becomes compressed to its minimum volume. Another way of saying this is that it never becomes "degenerate." Therefore, it cannot hold the star extended. Once the star began contracting, it seemed as if it would contract forever.

Chandrasekhar continued working on the problem, completing his thesis in 1933. He was elected fellow of Trinity College and remained at Cambridge for several years. During his stay he became acquainted with Eddington, who took an active interest in his work, visiting him almost daily. Chandrasekhar had considerable respect for Eddington, who by now was one of the giants of astronomy. His pioneering work on stellar interiors had made him world famous.

Chandrasekhar decided the best way to convince scientists of his idea of a critical mass was to work the theory out in detail. He completed the monumental task in 1934 and sent two brief papers to the Royal Astronomical Society. In January 1935 he was invited to give a talk.

He had considerable confidence now and was sure his work would be immediately accepted. But to his surprise, after he gave his talk, Eddington, who was completely familiar with his work, approached the podium to give a paper. Eddington began by defining and explaining the two types of degeneracy: ordinary degeneracy and relativistic degeneracy (degeneracy of relativistic electrons). He continued by saying, "I don't know whether I shall escape from this meeting alive, but the point of my paper is that there is no such thing as relativistic degeneracy."

Chandrasekhar was shocked—and angered. Eddington was obviously not convinced by his work, but rather than tell him privately, he was making a fool out of him publicly. Chandrasekhar wanted to counter, but realized that Eddington's prestige was so great that almost anything Eddington said would be taken as gospel. He left the meeting depressed. It almost seemed as if his career was coming to an end before it really got started—several years of hard work had been shot down in a single evening.

Strangely, Eddington's attack did not end with this meeting. He continued referring to Chandrasekhar's work as heresy, yet his arguments against it were vague and circular. Chandrasekhar could not understand them, and neither could anyone else.

Chandrasekhar later sent his work along with Eddington's rebuttal to Rosenfeld in Copenhagen who passed them on to Neils Bohr. Bohr was astounded by Eddington's nonsense, referring to it as rubbish. "Could you perhaps induce Eddington to make his views intelligible to human mortals," Rosenfeld wrote back.

Several other well-known scientists began to sympathize with Chandrasekhar but it was still many years before his work would be acknowledged. He wrote a detailed account of it and published it in a book titled *An Introduction to the Study of Stellar Structures.* He then left the study of white dwarfs.

Fortunately, Chandrasekhar's ideas did finally prevail, and astronomers eventually became convinced that there was a critical mass. But there was still the problem: what happened to stars that had a mass greater than this critical mass?

BEYOND THE WHITE DWARF

Let us return to our story of the life cycle of a star. Earlier we saw that our sun will ignite the helium in its core explosively, creating a "helium flash." Eventually, though, things will return to equilibrium and the helium will burn peacefully, leaving carbon and oxygen. But the carbon and oxygen will never burn because the temperature will not get high enough.

In more massive stars we have a different story. Consider one about ten times as massive as our sun. In this star there will be no helium explosion; when the helium is ignited it will burn peacefully, and the carbon that is left in the core will be ignited peacefully shortly thereafter when the temperature reaches about 3 billion degrees. This same cycle will continue through neon, magnesium, silicon, phosphorus, sulfur, and nickel, until finally the star's interior consists of numerous shells of nuclear burning, one within the other. This is, in fact, where most of the elements of the universe were generated.

But there is a hitch. When the star finally develops a core of iron, it is unable to go any further: iron does not burn. By the time the star has developed an iron core, though, the pressures and temperatures are so high that electrons and protons are being squashed into one another, leaving chargeless particles called neutrons. Neutrons take up much less room than electrons and the core begins to fall in on itself creating even more heat, thereby accelerating the process. Large numbers of particles called neutrinos are generated, which, unlike protons, can easily penetrate the outer layers of the star. Almost immediately they rush out through them and leave the star. This causes an energy vacuum in the core, which speeds up the infall even more. Within seconds the neutrino flux increases a millionfold, but the outer layers become much denser as they fall and the neutrinos cannot penetrate them; we would expect that they would therefore push them off into space. There is recent evidence, though, that it is the inner layers rebounding from the core that actually push them off. Anyway, within seconds, the star is blown apart by an explosion of incredible dimension—a *supernova*.

Besides distributing the heavy elements that the star has already generated into space, the supernova performs another important function. We saw earlier that elements up to iron are generated in burning cycles. But what about heavier elements such as silver, gold, and uranium? According to present theories, they are created in the supernova explosion itself.

There is still an important unanswered question: what is left after the outer layers are blown off? Fritz Zwicky of Mt. Wilson Observatory speculated in 1933 that a tiny star composed of neutrons, called a neutron star, would remain. A few years earlier, he and Walter Baade, also of Mt. Wilson Observatory, had begun an extensive study of supernovae. Since there were so few in our galaxy (only about one occurs every 50 years) they decided to look for them in other galaxies. Monitoring 3000 galaxies they observed 12 in a period of about 3 years.

Zwicky's suggestion that neutron stars might exist was, without a doubt, daring. Unfortunately, few people paid any attention to it. The Russian physicist Lev Landau worked out some of the details of such a star. Then in 1939 J. Robert Oppenheimer of the University of California became interested in the problem. He decided to apply Einstein's general theory of relativity to see if such objects could exist, assigning the problem to his student George Volkoff. Volkoff soon discovered that if a star was sufficiently massive its collapse would indeed lead to a neutron star. But even stranger he found, as Chandrasekhar had earlier in the case of white dwarfs, that there was a limiting mass.

When the electrons and protons came together to produce neutrons, the neutrons took up much less room than the electrons so that a denser object resulted. And again they showed that there was a degeneracy pressure as there was in the case of white dwarfs, only in this case it was neutron degeneracy pressure and it could hold up larger masses—up to about 3.2 solar masses. And again it left scientists speculating: what happens to stars with masses greater than 3.2 solar masses?

Most astronomers did not lose any sleep over the problem, though, because nobody had yet detected a neutron star. Did such a strange object actually exist? Many years passed before proof came that they did.

In the early 1960s Antony Hewish of Cambridge developed a technique for distinguishing narrow-angle radio sources in the sky from more extended ones such as galaxies. A few years earlier, starlike objects called quasars that emitted narrow-angle radiation had been discovered and Hewish thought his technique would be useful in locating them. He would, however, need a different type of radio telescope—one that was sensitive to sudden changes in frequency (most radio telescopes at this time were not). He decided therefore to build one and proceeded, with the help of several students, to cover a 4½-acre field with poles and wires.

One of the students was a doctoral candidate, Jocelyn Bell. When the project was finally finished in July 1967, she was assigned the task of analyzing the miles of recordings that the telescope generated. One of her duties was to identify any man-made interference that the telescope picked up, and within a few weeks she discovered a tracing that looked man-made—yet it was somehow different. When it continued to appear at approximately the same time each night, she became particularly interested in it and mentioned it to Hewish. He suggested she get a high-speed recording of it so its detailed structure could be seen, but by the time she was ready it had disappeared. She continued to watch for it for weeks, then finally gave up one day and went to a lecture at Cambridge. When she returned to analyze the charts—sure enough—it was there. The next day she managed to get the first high-speed tracing of it, and was surprised to find that it was a series of very evenly spaced peaks— one every 1.3 seconds. When she phoned the information to Hewish he replied, "Well, that settles it—it has to be man-made." What he realized, and she did not, was that it was almost impossible for an astronomical object to pulse that fast, with the exception of perhaps a white dwarf, or a neutron star, if they existed.

The official announcement of the discovery came in January 1968 and the news astounded the astronomical world. Some even suggested it might be a message from extraterrestrial beings. Theoreticians soon got into the act, their interest centering first around the possibility that the object was a white dwarf. But calculations soon showed that they could not be white dwarfs (assuming they pulsed)—the range of possible periods was too long. Neutron stars, on the other hand, pulsed too rapidly. The name they had been given, "pulsar," obviously was not appropriate; they could not be pulsating stars.

Another possibility was a lighthouse model. Perhaps they emitted a beam of radiation, or maybe two beams. As they rotated the beams would sweep by the earth, just as a beam from a lighthouse sweeps by a ship at sea. We would expect to observe a pulse every time the beam swept across the earth. This seemed to be a reasonable model, and the best candidate, as long as the period of the pulses was not too fast, was a rotating white dwarf.

But then a pulsar was discovered in the Crab nebula that pulsed at the rate of 30 pulses per second. White dwarfs could not rotate this fast; they would break apart under the stress. This left only rotating neutron stars. Tommy Gold of Cornell had long advocated that they were better candidates. He made the appropriate calculations for the energy release of a neutron star rotating 30 times per second and compared it to the known

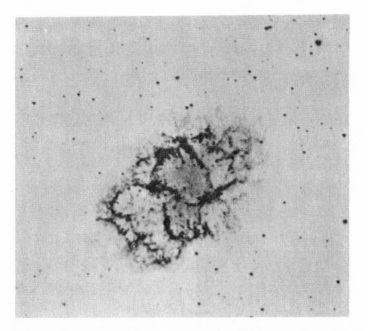

The Crab nebula. (© 1974 AURA, Inc. Courtesy National Optical Astronomy Observatories.)

energy release of the Crab. The two numbers were so close that there was little doubt: the Crab pulsar had to be a neutron star, and this meant that all pulsars were likely the same.

Within a short time details of a possible model were worked out. The spinning neutron star had an excessively strong magnetic field that spun with the star. Its strength was due to the collapse: even if the field was weak in the original star the collapse would "concentrate" it, making it exceedingly strong. Charged particles from the surface of the neutron star would move outward along the field lines generating electromagnetic waves (radio waves and visible radiation). A particularly important aspect of the model was that the magnetic field axis was not

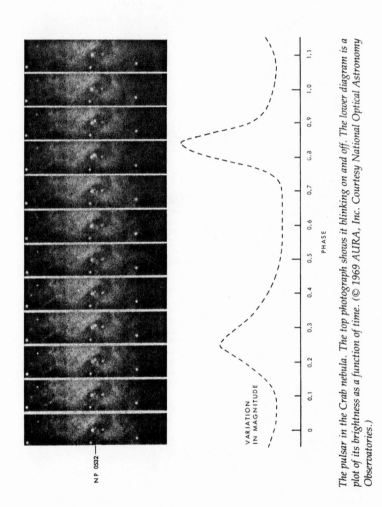

The pulsar in the Crab nebula. The top photograph shows it blinking on and off. The lower diagram is a plot of its brightness as a function of time. (© 1969 AURA, Inc. Courtesy National Optical Astronomy Observatories.)

necessarily aligned with the spin axis. Since the radiation would be channeled out the north and south magnetic poles of the star, and if this direction was, say, perpendicular to the spin direction, the beam would be spun around in the same way as a lighthouse beacon. If we happened to lie in its path, we would detect a flash of electromagnetic radiation.

The neutron star itself would be 10 to 20 miles across. It would have an extremely rigid surface—millions of times as rigid as steel—and beneath this surface would be what we call a superfluid: a mixture of neutrons and other particles. There may be a small core at the center.

It was noticed soon after the discovery of pulsars that they were gradually (very gradually) slowing down. In other words, their period was increasing by about a millionth of a second a month. This would be expected if they were releasing energy to space—as we knew they were. Unexpected, however, was the discovery that in some of them the period suddenly jumped. Astronomers called these sudden jumps "glitches." We now know that at least in the case of the Crab nebula this is due to a starquake. As the star slows down, its surface, which is quite oblate because of the high spin, relaxes back to its original form causing a slight crack in the surface.

REFLECTION

Now that we understand how a star evolves, let us stop and look back at the question posed at the beginning of the chapter. Where does general relativity break down? In other words, where does the theory become inadequate so that another (as yet undiscovered) theory is required? In asking this same question of Newton's classical theory, we saw that it broke down when we tried to apply it to atoms; a new theory—quantum theory—was needed. Quantum theory also breaks down at exceedingly high speeds and must be supplemented by special relativity.

It turns out that we do not even need general relativity in dealing with ordinary stars; Newton's theory gives us a satisfactory answer, so it obviously does not break down. It was important, though, to discuss the formation and life cycle of a star. It gave us the background we needed for understanding white dwarfs and neutron stars.

In dealing with white dwarfs we saw that the breakthrough came when Chandrasekhar applied both quantum theory and special relativity. Without them white dwarfs would be incomprehensible so it is obvious that Newton's theory is not adequate for these objects. General relativity, however, is again not needed to understand white dwarfs.

Then we came to neutron stars, much denser than white dwarfs. The first detailed calculations were done by Oppenheimer and Volkoff using general relativity, and indeed we find that this is the critical point: neutron stars or anything more dense cannot be understood or explained unless we use general relativity.

So far, though, general relativity is adequate (it does not break down). But what lies beyond neutron stars? As in the case of white dwarfs there is another limit. Neutron degeneracy pressure will support stars of up to about 3.2 solar masses. If a star collapses leaving a mass greater than this, we will see in the next chapter that we get an exceedingly strange object—a black hole. This is where general relativity begins to break down. But black holes are important in relation to another problem; we will see that they also give us the first link between quantum theory and general relativity.

CHAPTER 5

The Ultimate Abyss: The Black Hole

Astronomers have recently focused their attention on one of the most bizarre discoveries ever made. According to general relativity, there are objects in space that possess such powerful gravitational fields that when stars, planets, asteroids, or literally anything is pulled into them, it is crushed beyond recognition. Even stranger, once it passes through their surface there is no way it can ever emerge; it is forever cut off from our universe. We call such objects black holes.

Although there has been a lot of excitement about black holes in the last few years, the concept is not new; it has been around for over 200 years. The astronomer John Mitchell, Rector of Thornhill in Yorkshire, England, showed in 1784 that if the mass of a star were sufficiently great the star would trap its own light and be invisible to us. The French scientist Pierre Laplace arrived at the same conclusion a few years later.

To understand their argument, let us begin with what we call escape velocity. Consider several rocket ships blasting off from the earth, each with a greater velocity than the preceding one. The first couple may trace out arcs above the earth and fall back to its surface but eventually one will go into orbit. Then finally a velocity will be reached where they no longer orbit the earth, but escape to space. They will have, in effect, completely overcome the gravitational pull of the earth. This velocity is called the escape velocity, and it applies not just to rocket ships, but to anything (e.g., moons, particles). In the case of the earth, the escape velocity is approximately 25,000 miles per hour. More massive objects, though, have higher escape velocities; in fact, the greater the mass, the higher this velocity, which means that we will eventually reach a mass where the escape velocity is greater than that of light. If this happened in the case of a star, light could not leave its surface. This was the type of object visualized by Mitchell and Laplace. It is, in a sense, a black hole, but it is different from the ones we will be talking about.

Mitchell arrived at his "black hole" using Newton's theory but if we try to examine this object further using his theory we find it tells us almost nothing; for details we must turn to general relativity. The first person to realize that general relativity also predicted strange objects was Karl Schwarzschild. We saw earlier that he obtained a solution to Einstein's equation shortly after Einstein published it, but the solution bothered him. Mass curved space as Einstein had predicted, but the curvature became infinite at a finite radius—not at a point as might be expected. Space, in effect, closed in on itself cutting off a small region from the outside universe.

Schwarzschild communicated his solution to Einstein, who was pleased that a solution had been discovered, but was also troubled by the strange result. Einstein's attentions turned shortly thereafter to the unification of the gravitational and electromagnetic fields and it was during this investigation that he discovered something that disturbed him even more. Many scientists had begun to think that the fundamental particles of nature (i.e., electrons, protons) were associated with mathematical singularities (a mathematical singularity occurs when a mathematical expression becomes infinite). Einstein was examining these singularities with Nathan Rosen when he made a startling discovery: Instead of the usual single set of solutions to his equations, there were two sets. The first showed that the space leading to the singularity developed a long narrow "throat" but strangely, the second set also corresponded to a

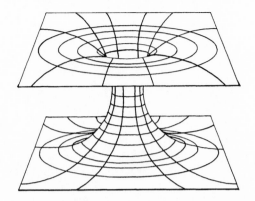

A wormhole in space, also called an Einstein–Rosen bridge.

throat that was attached to the other end of the first throat. And since this also applied to black holes, it meant that if you followed the throat far enough (i.e., to the black hole itself and beyond), it eventually began to open up again. But what did it open up into? The only answer seemed to be another universe. Einstein did not like this possibility and even today many scientists still feel uneasy when the discussion turns to other universes. These throats were soon referred to as Einstein–Rosen bridges; they are now sometimes referred to as "space-time tunnels." It was later shown that they need not lead to other uni-

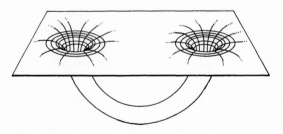

A wormhole joining two distant regions of space.

verses, but could also open into distant regions in our own universe. They were, indeed, interstellar subways.

Soon after Einstein discovered these bridges, he began to wonder if it was possible for someone to travel through them to these other universes. He was relieved when he discovered that a velocity greater than light was needed, and according to special relativity matter cannot travel that fast.

OPPENHEIMER AND CONTINUING COLLAPSE

About this time scientists began considering the possibility that collapsed stars would lead to black holes. In 1939, shortly after finishing his investigation of neutron stars with George Volkoff, Robert Oppenheimer turned his attention to stars too massive to end as neutron stars. Along with his student Hartland Snyder, he found that when the thermonuclear furnace of such a massive star went out there was nothing to stop its collapse and it therefore went on forever. The strange result confused Oppenheimer and his students, but unfortunately they never followed up on it.

Oppenheimer was born in New York City in 1904. His parents were fairly well-to-do and he was attended to by servants and maids. Because of this he was, at times, demanding and abrasive. But these traits did not impede his outstanding intellect even at a young age. His friends remembered him as a handsome lad with a mop of black hair and piercing blue eyes, handsome enough perhaps to have attracted many girls. But girls played a small part in his life when he was young; he was too busy studying most of the time. He was always at the top of his class, and his ego probably would have been shattered if he had ever ranked lower. His appetite for knowledge was so great, in fact, that it left little time for anything else—including sports, which he generally disliked anyway.

His scientific interests developed early. When he was about 6 his grandfather gave him a small collection of minerals, and he

J. Robert Oppenheimer (1904–1967), July 1962. (Courtesy CERN/AIP Niels Bohr Library.)

soon became an ardent collector. He was introduced to physics and chemistry in junior high and they fascinated him; part of the fascination was no doubt due to one of his teachers, Augustus Klock, for whom he had a great affection. Another teacher also had considerable influence on him—an English teacher in high school. Oppenheimer was among a group of students that was frequently invited to his house.

The friendship of these teachers helped but Oppenheimer was still considered snobbish even in high school. He liked to show off his knowledge and it occasionally got him into trouble. Because of his arrogance he was the brunt of a considerable amount of teasing. Once, while at a camp, he was locked up nude in an icehouse overnight because he had written to his parents that he was glad he had come because he was learning the facts of life. They became alarmed and rushed to the camp— and there was soon a clamp on all dirty stories.

He was described as frail and a little clumsy by most people who remembered him when he was young but a visit to the Pecos Valley of New Mexico while he was a teenager changed him. He developed an enthusiasm for trail riding and camping. He loved to sleep on the ground and occasionally went without food for long periods of time. Later he spent a considerable amount of time in this valley.

In 1922 he entered Harvard University as a freshman. He was still not certain what he wanted to become but because of his interest in chemistry he signed up for it, completing the usual 4-year program in 3 years. His colleagues all seem to agree that he had a natural ability beyond theirs.

During his 3 years at Harvard he never took a girl out. But he was not beyond innocent horseplay. One day while he and two friends were walking along the edge of a lake in the dead of winter, one of them dared him to go swimming in the nude. He replied, "I will if you will," and soon all three of them were skinny-dipping in the freezing water.

Academically, he continued with his usual string of As, but he did get an occasional B. He was uncertain about what he wanted to do until his third year when he took a course in thermodynamics given by Percy Bridgeman—his first course in physics. He thoroughly enjoyed it and was attracted to Bridgeman as a teacher, later commenting that he found him to be the most brilliant teacher of his Harvard years. He decided he wanted to go into physics instead of chemistry and upon graduation applied to Cambridge University in England. He wanted to work under Rutherford, who was one of the world's leading physicists, and although Bridgeman wrote an excellent letter of reference, Rutherford was not impressed. Instead of taking him in his own laboratory, Rutherford arranged for him to work under G. P. Thomson.

When he arrived in England, Thomson put him in a corner of his basement laboratory and assigned him the task of making thin films of beryllium. But Oppenheimer's inexperience in the lab proved to be his downfall; he did not like the work and made little progress. In addition to developing a distaste for lab work, he also disliked his living quarters, later referring to them as a "hellhole." This was not what he had expected and depression soon began to overcome him. He tried to ignore it, but it was soon evident to his friends that he was experiencing considerable inner turmoil. Some of his problems were overcome during a break when he and two companions made a trip to Corsica. They roamed the mountains and lived a carefree life for a couple of weeks. The real change in his attitude came, though, when Bohr visited his lab. He was thoroughly impressed with Bohr and the work he was doing, and immediately made up his mind to become a theoretical physicist.

He was invited by Max Born to visit the University of Göttingen in Germany shortly thereafter and he left Cambridge, thinking at first that it would be only a short visit, but he enjoyed the intellectual atmosphere and attitude at Göttingen so much he stayed on and got his Ph.D. there a few years later. A significant change in his life occurred in Göttingen; he shed any traces of depression and immersed himself completely in his work. The years he was in Germany (1926–1927) were important ones in physics; the new theory of "quantum mechanics" was just developing and significant breakthroughs were being made almost weekly.

Oppenheimer was right in the midst of the developments and it was an exciting time for him. He collaborated with Born, using the new theory to explain the collisions of particles; they eventually created what is now referred to as the Born–

Oppenheimer approximation. Paul Dirac also soon came to Göttingen from Cambridge, and he and Oppenheimer, already good friends, spent much time together talking about the new developments.

In 1928 Oppenheimer sailed back to the United States. He was now on top of the world: he had participated in the development of quantum theory and had acquired a reputation, and was therefore in considerable demand. Ten universities in the United States and several in Europe were after his services as a teacher. Even his old alma mater, Harvard, wanted him. After thinking it over he decided upon a joint position at Berkeley and Caltech at Pasadena. Berkeley particularly appealed to him because it had an excellent faculty but was a desert in theoretical physics. Furthermore, no one knew anything about quantum theory.

But he was not an immediate success as a teacher at Berkeley. According to his first students, he raced through each lecture in a series of mumbles, always with a cigarette in his hand. He had a cigarette in one hand and chalk in the other so often that the students began betting on the possibility he would attempt to smoke the chalk, or write on the board with the cigarette. But apparently he never did.

He soon began to realize, though, that the students were not following him and he slowed down, and as he did his classroom technique improved. His forte was perhaps the time he spent with students outside the classroom. He would gladly spend hours explaining things or trying to get a point across, and he generally praised his students even when they did not deserve it. But in class he could be ruthless, and many were slightly terrorized by him. His graduate students—and there were usually at least a dozen—came to idolize and even emulate him. He took them to cafes, introduced them to new food and wine, and spent much of his free time with them. They were eventually referred to jokingly as "Oppie's cronies." But they were the cream of the crop, some of the best in America. Many of them went on to become famous in their own right.

Oppenheimer was particularly productive during this peri-od of his life. He published many papers, most of them impor-tant, yet none outstanding enough to place him in running for the Nobel prize. One of his students once said that although he had a sharp mind and penetrating intellect, he never reached the top levels primarily because he never followed up on his developments.

About 1940 a new phase in his life began. In 1938 nuclear fission had been discovered by Hahn and Strassman in Ger-many, and it soon became evident that a powerful bomb could be built. Because the United States was now at war with Ger-many, and because it was known that the Germans were in the process of building such a bomb, the United States threw itself into the project. Oppenheimer was in on most of the initial meetings. At one of the early ones he presented a calculation on the amount of fissionable material needed for such a bomb, and in 1940 he was named administrative head of the project. His first job was to find a site for a secret laboratory where the bomb could be assembled. Familiarity with the New Mexico moun-tains led him to select Los Alamos. And, needless to say, the bomb was built, and used.

Oppenheimer's major contribution to astrophysics was his discovery of the ever-collapsing star. This, interestingly, was his only excursion into the area; most of his work during 1938–1939 (when he published this paper) was on quantum theory and nuclear physics. The strange result he got amazed him; writing to another scientist in 1939 he stated, "We have been working on static and nonstatic solutions for very heavy masses that have exhausted their nuclear energy sources; old stars perhaps which collapse to neutron cores. The results have been very odd. . . ."

The ever-collapsing star discovered by Oppenheimer and Snyder attracted some attention for a while, but gradually in-terest waned. Amazing results, thought many astronomers, but of little consequence in present-day astronomy. How could such exotic objects possibly exist in nature? Oppenheimer would no doubt have continued his work on collapsing stars but the war intervened and his attentions turned to the atomic bomb. He never returned to the problem.

Little was accomplished during the 1950s, with barely a handful of scientists around the world working on the problem. But suddenly in the early 1960s things began to change. Strange starlike objects (now called quasars) were discovered, and when their energy was measured it was found to be enormous. How could they be generating so much energy? Could black holes be involved? Interest was generated, but neutron stars had not even been found so the black hole concept was not taken very seriously.

Then in 1967 the first pulsing signal was picked up by Jocelyn Bell and within a year astronomers were confident they had detected the first neutron star. Their attention soon turned to black holes; they had been known to exist on paper for years but did they actually exist in nature? Once scientists became interested in black holes, theoretical progress was rapid. John Wheeler, Kip Thorne, Remo Ruffini, and others were soon hard at work in the United States, as were Y. B. Zeldovich and I. D. Novikov in Russia, and Roger Penrose, Brandon Carter, and Stephen Hawking in England. Within a few years the theory of black holes was well developed.

The fundamental tool in the study of black holes is general relativity, but it should be pointed out that black holes are not an outgrowth of general relativity. If, for example, sometime in the future general relativity is proven to be incorrect, this does not mean the end of black holes. They exist in all seriously considered theories of gravity. A theory put forward a few years ago by Dicke and Brans, now considered to be general relativity's most serious competitor, also predicts black holes.

From a distance there is little that is exotic about a black hole, except perhaps its eerie appearance. Its gravitational field is the same as it was before the collapse. If a planet revolved around a massive star, and the star suddenly collapsed to a black hole, the planet would remain in orbit. It would, in fact, continue revolving around it for perhaps billions of years. Eventually, though, as a result of certain forces it would slowly approach the black hole, and as it passed a critical point it would be drawn in and crushed.

But if the gravitational field is the same after the collapse as it was before, what is all the fuss about overwhelming gravitational forces? It is true that the field around the star does not change but it is important to remember that the original star was perhaps a million miles across; the black hole, on the other hand, may be only 10 miles across. This means you can approach much closer to the source of the field, and as you do its strength increases.

No light is emitted from a black hole, yet if you approached one in a spaceship you would know it was there. You would feel its gravitational field, but aside from that if you got close enough you would be able to see it through a telescope. Actually, what you would see is a black circle that stands out against the background stars; no light comes from the black hole itself. You would have to be careful at this point, though; if your spaceship got too close it would be pulled in—and there would be no escape.

COLLAPSE OF A STAR TO A BLACK HOLE

To see some of the consequences of black holes, let us consider the collapse of a star sufficiently massive to give us one. We will assume that the star is not rotating. As it ages it uses up its fuel and bloats into a red giant, but finally its fuel is depleted and it becomes unstable. The thermonuclear furnace once provided an outward pressure to balance the inward pull of gravity, but it is now out. The tremendous inward force of gravity soon overwhelms the star. In the case of small stars (less than 1 solar mass), the collapse occurs slowly over millions of years, but for massive stars the collapse is almost instantaneous. The core of

How a black hole might look if you encountered one in space. The large number of stars near it is an optical illusion caused by the extreme curvature of space around it.

the star falls in on itself in less than a thousandth of a second and becomes a black hole.

Let us assume now that we slow the collapse down and watch it in slow motion. Soon after it begins, surges of X rays and γ rays are given off. But as the collapse continues, the photons find it increasingly difficult to fight the ever-increasing gravitational pull. Those that leave the surface at an angle are forced into curved paths (as predicted by general relativity). Finally, those that attempt to leave the star parallel to its surface are pulled into orbit around it, then a fraction of a second later no photons at all can escape; the star has passed through what we call its *event horizon*. We can no longer see it directly—we see only a black sphere in space. But the matter of the star continues to collapse beyond the event horizon; in fact, it continues to collapse forever and is eventually crushed to zero volume at the center of the star. This center point is called the singularity.

Has anyone ever seen a star collapse in this way? The answer to this has to be no—for several reasons. First, the star collapses too fast. We would just see a giant star at some point in space, then suddenly it would disappear—assuming we were

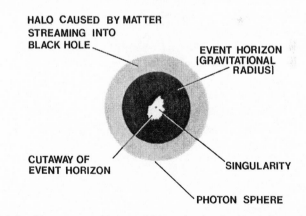

HALO CAUSED BY MATTER
STREAMING INTO
BLACK HOLE

EVENT HORIZON
(GRAVITATIONAL
RADIUS)

CUTAWAY OF
EVENT HORIZON

SINGULARITY

PHOTON SPHERE

A black hole with a cutaway of its event horizon showing the singularity at the center.

lucky enough to catch the collapse. But, of course, even this is unlikely: in our lifetime only a few stars in our neighborhood of space have become black holes.

Let us return to the collapse and look at it a little more closely. Again, if we could see it in slow motion we would see it shrink in size and turn red. The redness is a result of the slowing of time predicted by general relativity. Photons are like little clocks, vibrating with a very precise frequency; if time slows down, this frequency decreases, causing them to redden.

As the star nears its event horizon the emitted light is trapped in orbit around it, creating a reddish halo that lingers for a while. But finally the red glow fades and we see only a black sphere—the black hole.

INSIDE A BLACK HOLE

So far we have seen the collapse only through the eyes of a distant observer. He saw the star decrease in size until finally when it became black it decreased no further—in effect, it froze at a certain size. A closer look would show us that the star is approaching a critical size; it gets smaller and smaller but somehow never seems to reach this critical size. But what about an observer who is on the surface of the star as it collapses? Would he see the same thing? It turns out that he would not—things would be quite different for him. The star would appear to collapse in a finite time; in a fraction of second he would pass through the event horizon and be crushed at the center along with the mass of the star. Yet, according to the distant observer, he would still be on the surface of the collapsing star, even years after the collapse began.

This seemingly odd result arises because of the strange way time behaves: it passes at a different rate depending on how close you are to the black hole. Suppose we have two observers, A and B, at some distance from the black hole; each has a watch and they synchronize them. One of the observers, say B, waves goodbye to the other and projects himself toward the black hole.

The stationary observer (A) watches him fall closer and closer to it. He observes B's watch through a telescope and sees that it runs slower and slower as he approaches the hole. Finally it almost stops, yet somehow it never quite stops—just as he never quite seems to reach the surface of the black hole.

Now let us consider what the falling observer, B, sees. For him there is no frozen star; he falls rapidly toward the black hole and if he looks at his watch time appears to be running normally. If he looks back at A's watch, though, he sees that it is running fast. In fact, the closer he gets to the black hole the faster it runs.

He also notices something else as he closes in on the black hole: he is being stretched; something is pulling him apart. This is caused by what we call tidal forces; they act whenever there is a large change in gravity over a small distance. Assuming his feet are closer to the black hole than his head, they will be pulled toward it with a greater force; his body will therefore be stretched. The same effect occurs to a lesser degree when you approach a neutron star. Just before our observer reaches the black hole, his body will likely resemble a piece of string. We will see later, however, that if the black hole is sufficiently massive, the tidal forces are small; we will assume this is the case here so we can finish our story.

In a finite (but very short) time by his watch the falling observer passes through the event horizon and enters the never-never land within. Once he has passed this horizon he is forever cut off from the outside world. He can never return to it, nor communicate with it in any way. That is, in fact, why we call it an event horizon; it represents the end (the horizon) of events in our universe.

Once he is inside the black hole, our observer cannot report back to us what he sees; he is drawn ever closer to the center of the black hole. If he tried to race back to the event horizon, he would find it receding from him at the speed of light, and of course he cannot travel that fast. At the center of the sphere is the collapsed remains of the star—the singularity.

As the observer falls toward this singularity he soon notices that space and time have, in effect, interchanged their roles. On the outside of the event horizon we have control over space, but no control over time—it passes in the same way regardless of what we do. But strangely, inside the event horizon we have some control over time, but none over space; we are drawn ever closer to the singularity regardless of what we do. And when we get to it our fate will be the same as the star's—we will be squashed to zero volume.

OTHER TYPES OF BLACK HOLES

The black hole discussed above was nonrotating. Since the solution of Einstein's equations corresponding to it was discovered by Schwarzschild, it is referred to as a Schwarzschild black hole. But most, if not all, stars spin and therefore the black holes that form from them will be spinning. The solution for this case was discovered by Roy Kerr of the University of Texas in 1963. It was a much more complicated solution than Schwarzschild's, and the black hole was correspondingly more complex.

If you approach a Kerr black hole the first thing you would notice is that you are being pulled around it in the direction of its spin. And the closer you approach the faster you would be spun. At a certain distance from the spin axis you would find, in fact, that you are being spun at close to the speed of light. The surface where you would be moving at the speed of light is referred to as the static limit. If you venture inside of it, you will find that this black hole also has an event horizon, and as in the case of the Schwarzschild black hole it is spherical in shape. The static limit, on the other hand, is oblate, the two surfaces touching only at the poles. The region between the surfaces is called the ergosphere.

Passing inside the event horizon you find there is also a singularity, but it is different from the previous one—it is in the form of a ring. There is another important difference in this case.

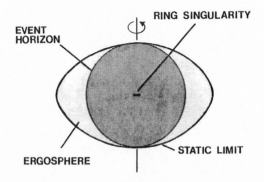

A Kerr black hole.

Einstein showed in the case of the Schwarzschild black hole that you need a velocity greater than that of light to pass through the associated wormhole. In this case you could pass through with a velocity less than that of light.

Let us turn now to a more detailed look at the collapse of a spinning star. First of all, we know that if a star is spinning it will spin faster as it collapses because of conservation principles. A skater uses this principle to increase her spin: she begins with her arms extended, and as she pulls them into her chest she spins faster. In the case of a collapsing star, even if the spin is moderate, such as it is in the case of our sun, it will have increased so much by the time it has collapsed it will fly apart before it becomes a black hole. To end up as a black hole a star must get rid of some of its spin, and it seems likely that most do. Assuming that they do, it is reasonable that most massive stars end as Kerr black holes.

There are, however, two other types of black holes. They may not occur in nature but they are important theoretically nevertheless. When a star becomes a black hole, almost everything associated with it is lost into the singularity. We could never, for example, know its exact composition or temperature; these things were lost when it became a black hole. Only three properties remain: mass, rotation, and charge. Because of this there are four types of black holes. Besides the Schwarzschild and Kerr, there is the Reissner–Nordström black hole (nonspinning with charge) and the Kerr–Newman black hole (spinning with charge).

In 1971 the British theorist Roger Penrose proved that energy could be extracted from black holes with spin and/or charge. If a pellet, for example, were projected into the ergosphere, broke apart in this region with one part falling into the event horizon and the other emerging, the emitted part would emerge with considerably more energy than the original pellet entered with. Energy would, in effect, be extracted from the black hole. In the case of the Kerr black hole, this energy loss would show up as a decrease in its spin.

SEARCHING FOR BLACK HOLES

So far we have been talking about black holes from a theoretical point of view, but do they actually exist in nature? This was a question uppermost in the minds of astronomers in the mid-1960s. Many were skeptical, and a few are still skeptical even today. General relativity is, after all, only a theory, although most scientists feel that it is an excellent theory and are certain its predictions are correct. It is possible, though, that it may some day be superseded and the type of black hole it predicts will not exist in the new theory. It was important, therefore, that a good candidate be found. But where would we look? Or for that matter would they be worth looking for?—there might be so few of them in our galaxy that it would be unlikely we could ever locate one. Let us begin by considering this last point: How many black holes are there likely to be in our galaxy?

Time is, of course, an important factor. Has enough time elapsed for a large number of black holes to have developed? We know our sun has a lifetime of about 10 billion years and it is now about 4.5 billion years old. But black holes result from stars that are much more massive than our sun, and these stars evolve much faster. Most massive stars live out their entire lives in less than a billion years. Time, then, appears to be on our side.

Next, we must consider the number of massive stars in our galaxy. Is it sufficient? A final mass of only 3 solar masses is all that is needed for a star to end as a black hole. But most stars lose mass before and during their final collapses and therefore a black hole of 3 solar masses probably resulted from a star that initially had considerably more mass than this, possibly as much as 8 solar masses. Fortunately, even this is not excessive; many of the stars in our galaxy are this massive.

Our galaxy contains about 200 billion stars and is approximately 15–16 billion years old. How many black holes is it likely to contain? There are a large number of uncertainties in trying to make an estimate and it will be crude at best. We will begin by assuming one black hole forms in our galaxy every 100 years. This is based on what we know about the distribution of stars in it, and their life cycle. With this we can easily calculate the total number; the calculation gives several hundred thousand, which again may be off by several factors, but at least it makes us feel confident that the search is worthwhile.

The next question is: What do we look for? Since most black holes are only a few miles across, it is unlikely that we would be able to see one directly. We obviously must resort to indirect methods. The best of these centers around the effects that the black hole has on gas that falls into it. When gas spirals into a black hole, it is heated so much that it gives off X rays, X rays that we should be able to detect from earth.

Let us consider a scenario, then, in which X rays are generated. Assume we are dealing with a double star system where one of the stars has just collapsed to a black hole. If gas from the other star is somehow pulled into the black hole, X rays will be generated. How could this happen? To answer this we must consider what is called the Roche lobe. Around the black hole there are a set of imaginary spheres where the gravitational field is the same at all points on a given sphere; they are called potential spheres. When we have a binary (double) system, the spheres around the individual stars become distorted because the gravitational field of one of the stars affects the other. There will be a particular pair, though, in the form of a figure eight where the field strength is the same at all points along the figure:

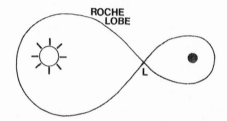

The Roche lobe around a star. L is the Lagrangian point.

This is the Roche lobe. Such a lobe exists for all double systems, even the earth and moon. The crossover point (the point where the two lobes join) is particularly important; it is called the Lagrangian point (L). If any matter from star A passes this point, it will be pulled into B, and vice versa.

Suppose now that B becomes a black hole. This means that if any gas from A passes L, it will be dragged into the black hole. There are two ways this might happen. First, we know that if we wait long enough, star A will begin to expand—it will become a red giant—and its outer layers will pass L. Another possibility is that star A is a large blue star that has an extensive solar wind. The particles that constitute this wind will also be pulled into the black hole if they pass L.

According to calculations, any material that passes L will spiral in toward B, forming what is called an accretion disk. The individual particles of gas in this disk will behave like the planets of our solar system in that those closer to the black hole will move faster than those farther out (just as Mercury travels faster in its orbit than the earth does). This creates considerable friction between the layers, which heats the gas. Just before it enters the black hole it will reach billions of degrees, resulting in the creation of intense X rays.

Binary systems are therefore prime candidates in our search, but since black holes are extremely small, any that are present in such systems would be invisible. Could we detect a system such as this? Astronomers are, in fact, familiar with such systems; they are called spectroscopic binaries. Although they see only one star through the telescope, they know there are actually two because of the behavior of the spectral lines of the star: they move back and forth because the wavelength of the light emitted is altered by its motion (called the Doppler effect).

Before discussing the X-ray sources of this type that were eventually found, let us briefly outline the history of X-ray astronomy. Our atmosphere cuts off X rays, and because of this, rockets, balloons, and satellites are needed to get above it. The first rocket containing X-ray detection equipment was launched in 1962. Several X-ray sources were found immediately; one in the constellation Scorpius was later shown to be associated with a bright blue star, but there was no evidence that it was in a binary system. In the years that followed, other rocket flights were made and other sources were found. Two particularly interesting ones were found in the constellations Centaurus and Hercules, respectively. Both pulsed rapidly and both appeared to be in a binary system, but neither appeared to be associated with a black hole.

With the launching of the first X-ray satellite UHURU (the Swahili word for freedom) from Kenya on its independence day in December 1970, X-ray astronomy developed rapidly. The first UHURU catalogue of X-ray objects was soon published; it contained over 100 entries, 55 which were of particular interest (the source of their X-rays was unknown). Interest soon began to center around one known as CYG X1 in the constellation Cygnus. It pulsed rapidly but its pulses were unlike those of the ones in Hercules and Centaurus; they were not periodic. The shortest pulse times indicated the source was small—black hole-sized. Finally, in 1971 the optical component of the system was found (it is well beyond the limit of the naked eye). The system was indeed a spectroscopic binary with a period of 5.6 days, and the secondary (the X-ray source) could not be seen. The primary was a blue giant star known as HD226868 in the Henry Draper catalogue.

What were the masses of the two components? If the spectral class of the primary could be determined, we could calculate its approximate mass. It turned out to be about 22 times that of our sun. Using this and making various assumptions, the mass of the secondary was shown to be about 8 solar masses—easily enough to be a black hole. In short, we had a source of X rays associated with an invisible and therefore small object 8 times as massive as our sun—all the requirements needed for a black hole. Because the primary is a blue giant, it is assumed that the solar wind, and not the outer layer of the star itself, is being trapped and whirled into the black hole.

OTHER BLACK HOLE CANDIDATES

CYG X1 is without a doubt our best candidate, but it is not our only one. A similar source in the constellation Circinius called CIR X1 has been attracting considerable attention lately. It is a spectroscopic binary with an orbital period of 16.6 days. Unlike CYG X1, it has an off state, caused perhaps by the X-ray source being eclipsed by the primary. Time variations in the signal are just as short as those of CYG X1, indicating it is no bigger. The primary is a faint red star; some astronomers believe that its color and faintness arise because it is surrounded by a dust cloud; its light dims as it passes through the cloud. The basic difficulty with CIR X1 is that its mass has not been determined. In this respect it is not as good a candidate as CYG X1.

Another candidate lies about 5000 light-years away in the direction of the constellation Scorpius. In this case the mass is known; in fact, it is the most positive feature of the candidate.

The primary has a mass of 20 to 30 solar masses, which indicates the secondary (the black hole candidate) is from 7 to 11 solar masses. But as in the case of CIR X1 there is a difficulty: the time variations in the signal are not short enough to indicate it is black hole-sized.

The above candidates are all spectroscopic binaries, but we also have a candidate that does not belong to a binary system: Cas A. Cas A is an X-ray source that is believed to be the remnant of a supernova that occurred in approximately 1668. Strangely, there is no record of a supernova at that time. The Russian astrophysicist I. S. Shklovski, in studying Cas A, has found that it likely at one time had a mass about 20 times that of our sun. About half of this mass has been pushed off into space

The quasar 3C 273 in the constellation Virgo. (Courtesy National Optical Astronomy Observatories.)

in an expanding shell around it; the remaining central object now has a mass of about 10 solar masses. Shklovski believes that rather than explode as a supernova, most of the star imploded giving rise to a black hole. He feels he has sufficient proof to back up his claim.

Some of the most fascinating black hole candidates are not the result of stellar collapse (at least not directly), but rather those associated with groups of stars, or galaxies. The giant elliptical galaxy M87, which lies about 60 million light-years from us in the cluster of galaxies known as the Virgo cluster, is both a strong radio source and a strong source of X rays. Photographs of the object show a 4000-light-year-long jet emanating from the core. There appears to be considerable turbulence along the jet, and several large knots are clearly visible in it. The jet and the core of the galaxy are the strongest radio regions. Studies show that the stars near the core are moving at a tremendous rate (approximately 250 miles/sec) and there is evidence for gas inflow in the region. This indicates that large numbers of stars have accumulated in the core and it should be excessively bright. But it is not, yet it is massive—perhaps as massive as 5 billion suns. Many astronomers are convinced that this massive object is a black hole. Gas and stars may be whirling around it in a gigantic accretion ring, and as they are pulled in X rays are emitted.

M87 is not the only galaxy of this type in the sky; there are quasars and other galaxies that also have jets emanating from them. Some astronomers believe that all radio galaxies, and even perhaps ordinary galaxies such as ours, may have giant black holes at their core. We know, for example, that the core of our galaxy is an extremely strong radio source, but we are not yet certain what is causing it.

A REALISTIC LOOK AT THE EXOTIC POSSIBILITIES

Earlier we talked about the possibility that matter could pass through the space-time wormhole (Einstein–Rosen bridge) associated with a Kerr black hole. This implies, of course, that an astronaut could perhaps also pass through it—obviously an exciting possibility to contemplate. Before we look at the details of what his trip might be like, we must ask ourselves what the throat attached to the other end of the black hole (the one discovered by Einstein and Rosen) actually represents. It obviously cannot be associated directly with a black hole. Black holes only pull matter in; an astronaut would have to exit through this end so it would have to expel matter.

Astronomers refer to this end as a white hole. White holes are, in effect, time-reversed black holes, and because of this we would expect matter to come gushing out of them. Are there any objects of this type in nature? There is, indeed, evidence that matter is being expelled from the cores of compact galaxies called Seyferts; furthermore, quasars also appear to be expelling matter. It seems, therefore, at least on first sight, that white holes may have some significance.

If they do exist, it would mean that our astronaut could pass into the throat of a black hole, carefully avoid the singularity (which would crush him if he did not), then pass out through the throat associated with the white hole. All this seems easy enough—until we look at the details. First, there are the tidal forces that we mentioned earlier; they would tend to pull you apart before you ever got to the event horizon. Is there anything that can be done about them? It turns out that you can, at least in theory, get around them. The tidal forces associated with the usual stellar-collapse black hole (i.e., one a few miles across) are intense and there is no way you could get near them without being pulled apart. In a more massive black hole, however, these forces are less intense; in fact, the more massive it is, the weaker they are. If the black hole were excessively massive (millions of times as massive as our sun), they would be so weak they would not bother us as we passed through the event horizon.

But tidal forces are not the only thing we would have to overcome. Kerr black holes are spinning at a high rate and as our rocket ship approached it would be pulled around in the direction of spin. Furthermore, if there was matter in the vicinity, the radiation levels might be exceedingly high. It is, of course, quite possible that we could overcome these difficulties but there is another important point we have generally neglected: If we passed through the wormhole, where would we end up? According to Einstein, the only answer seemed to be "in another universe," but more recently it has been shown that there is another possibility: a distant point in our own universe. Somehow this seems more satisfying than the former possibility. It is possible, but not necessarily easy, to visualize our wormhole tunneling its way to some distant point in our universe, but as for tunnels to other universes—well, we do not even know what the term means, and most astronomers feel very uncomfortable discussing the concept.

The idea of wormholes as subways through space-time has, however, caught the fancy of many authors. They speculate that they may be our route to the stars. If we wanted to go to the star Alpha Centauri, for example, we would merely find a nearby entrance, pass through the subway, and exit somewhere near Alpha Centauri. There is, of course, a problem: we could never be sure where we would exit unless we had already passed through the wormhole. And the first time we went through, we would likely have considerable trouble trying to find out where we were relative to the sun. Another obvious problem with our subway is that it is one-way. When we got to our destination we would have the problem of getting back. Assuming we are not worried about that right now, let us try to visualize what such a trip might be like. As we approached the entrance we would be spun rapidly around it, but by blasting our rocket appropriately we could slow ourselves down. Finally, though, we would find that regardless of what we did we could not stay motionless— we are inside what is called the static limit. At this point we are whirling rapidly around in the ergosphere. As we continue moving inward we pass through the event horizon. Once inside we find that we have lost control of the spaceship; it is pulled rapidly toward the singularity regardless of what we do, and it appears as if our trip has been in vain. This occurs because space and time reverse their roles inside the event horizon. In our world we have control over space; in other words, we can move anywhere we want in space. But this is not true in the case of time; it passes regardless of what we do. Inside the black hole, however, where the two roles are reversed we have control over time, but no control over space. In other words, the space between us and the singularity always decreases regardless of what we do. Fortunately, scientists have shown that inside the usual event horizon there is a second event horizon, and when we pass through it there is another reversal of the roles of space and time, and they return to their usual roles. We can therefore avoid the singularity.

Assuming we have avoided the singularity, we can then pass out through the white hole throat and emerge at some distant point in our universe. If we look at our watches the whole trip may have taken only seconds. But if we could somehow compare our time to that back on earth, we would find that we may be emerging millions of years in the future, or perhaps in the past. Space-time wormholes are time machines!

H. G. Wells speculated about time machines many years ago. The ones he visualized were, of course, quite different than those associated with black holes. But here we are with "time machines" predicted by a respectable theory—and it is not too difficult to see that it leads to problems. Not simple ones, but ones complex enough that the whole idea of wormholes may have to be abandoned. One of the fundamental principles of physics is that of causality: for every effect there is a cause, and the cause must come before the effect. This principle may be violated if space-time wormholes exist.

If this were the only difficulty with wormholes, we might be able to accept them, but unfortunately there is more. Let us go back to the white hole; earlier I mentioned that it is a time-reversed black hole, and since black holes will exist for all future time, white holes must exist for all past time—in other words, since the beginning of the universe. But white holes associated with the stellar collapse of black holes could not possibly have existed this long. The star may have collapsed only a few years ago. There is, however, a type of black hole that we will talk about shortly called a primordial black hole. These are black holes that were presumably created in the big bang explosion, the explosion that created the universe. Assuming they exist, we would have a white hole (associated with a black hole) that has existed for all time.

But even this does not solve our problem completely. Doug Eardley of Yale University has shown that even if white holes were formed shortly after the big bang explosion, radiation would pile up around them and soon convert them into black holes. Thus, there may be no white holes in the universe; this would mean we would have entrances to our space-time subways but no exits. Once we entered one we might not be able to get out. Furthermore, John Wheeler and others have shown that these wormholes are extremely unstable. They might pulse rapidly—opening and closing in a way that would not allow us to pass through them.

I am sure, though, that all this will not deter science fiction writers. They will no doubt continue writing stories about such trips, but whether or not they will be possible—even in the very distant future—well, things look bleak. But of course, science does sometimes take strange twists.

WHERE THE THEORY BREAKS DOWN

I think this is enough speculation about black holes; our purpose in this book lies in a different direction: the possibility of an ultimate, all-encompassing, unified theory. But as we will see shortly this theory is closely tied to black holes. For now, though, we want to know where general relativity breaks down; we saw earlier that it is adequate when dealing with neutron stars and it predicts the existence of black holes.

Let us look closer, then, at what general relativity tells us about black holes. As we saw earlier, when a sufficiently massive star collapses it leaves an event horizon, which appears to us as a black sphere. The matter of the star, however, continues to collapse, creating finally a singularity at the center of the event horizon. Scientists usually describe this singularity as a place of infinite density and no dimensions. Yet somehow all the star, and anything that fell into the star later, is here. Because it is crushed to zero volume it has lost almost all its identity. We have to be careful, though, when we describe the singularity in this way; in reality, it is only a place where our theory breaks down. In other words, the place where conditions are such that we can no longer adequately describe what is really going on—and if we try an absurdity arises. The absurdity in this case is an infinite density and a mass of no dimensions. We say this is what occurs because of our ignorance, and this seems to be the logical end point, or perhaps the illogical end point, of the collapse.

Since the theory we are applying is general relativity, we finally have the answer to our question: general relativity breaks down, or is no longer valid, at the singularity of the black hole. It actually breaks down slightly before a true singularity occurs. If we try to apply it beyond this breakdown point, we get absurd results just as scientists did early in this century when they tried to apply classical (Maxwell's) theory to the atom. Maxwell's theory told us that atoms could not exist. The electrons whirling around the nucleus would radiate away the atom's energy, and within a short time they would collapse into it. This meant that all matter should be in a collapsed state—but it is not. Obviously something was wrong with the theory we were applying, or perhaps we were extending it into regions where it did not apply. Within a few years it was shown that, indeed, it could not be applied to atoms. Quantum theory was used and everything was explained satisfactorily.

Now we have general relativity giving us a seemingly absurd result. But Stephen Hawking and Roger Penrose proved a theorem a few years ago that told us that some sort of singularity—not necessarily the type we have been visualizing—always occurs inside an event horizon. More exactly, they showed that an "end or boundary" of space-time had to occur. We are still not sure what the significance of this is.

Let us go back to our collapsing star to see exactly what the difficulty is. The matter of the star continues to collapse after it passes inside the event horizon, eventually becoming so small that quantum effects become significant. Unfortunately, we do not have a quantized version of general relativity so we can only speculate about what probably happens when the matter gets into this domain. Scientists have speculated that quantum gravity may affect the fundamental nature of space itself. It may cease to be the continuum we know; it may break apart and become twisted and distorted in strange ways. It is, of course, difficult to visualize this breaking up—what, for example, is between the broken pieces of space? Leaving aside difficulties such as this, we can visualize large numbers of wormholes developing. The topology of the space would likely be extremely complicated at this stage—a kind of froth of space and non-space.

This is, in effect, the region we cannot describe with general relativity; a new theory is therefore needed to explain it. In the next chapter we will see that the same problem crops up in relation to the early universe.

THE DISCOVERY OF EVAPORATING BLACK HOLES

So far we have been talking about stellar-collapse black holes, i.e., black holes that are created in the collapse of a massive star. But there is another type. To see how it arises, consider the early moments of the big bang explosion. About 18 billion years ago all the mass of the universe was contained in an infinitely massive primordial nucleus—the universal singularity. This singularity suddenly became unstable and exploded, creating the universe.

One of the important questions in relation to this explosion is: Was it perfectly homogeneous or did inhomogeneities (clumping) develop in it? To answer this we merely have to look around us. The galaxy we live in and those around us tell us there had to be inhomogeneities; if not, these galaxies would not have developed. The universe would consist of a uniformly expanding gas. But if there were inhomogeneities, it is highly probable that pockets of matter were compressed as other matter expanded around them. These pockets may have been compressed into black holes, and unlike stellar-collapse black holes they would not all have a radius of a few miles. Some of them would be extremely small—smaller than protons—and others might be gigantic, with masses the order of those of galaxies. To distinguish these black holes from stellar-collapse ones, we refer to them as primordial black holes.

There has been considerable speculation about these primordial black holes over the years: suggestions that small ones—mini black holes—may have struck the earth, and suggestions that they might form the nuclei of strange heavy atoms. According to calculations, if one did strike the earth it would pass directly through with only a small explosion upon entering and another as it exited on the other side. We have no evidence, however, that this has ever happened.

One of the most significant discoveries in black hole physics in the last few years concerns these small black holes. Jacob Bekenstein was examining the thermodynamics of black holes in 1972 when he noticed that they seemed to have a surface temperature greater than 0°K (absolute zero—the lowest possible temperature in the universe). But this made no sense: everything that comes near a black hole is pulled in, nothing can get out. It therefore cannot emit anything, including radiation, and must have a temperature of 0°K. Bekenstein ignored the result, thinking it was some sort of mathematical anomaly. Stephen Hawking of Cambridge University, however, in examining it, showed that it was valid: black holes did have a temperature greater than 0°K. What Hawking did different from Bekenstein was apply quantum theory and show how the temperature arose. He wrote, "The paradox remained until 1974 when I was investigating what the behavior of matter in the vicinity of a black hole would be according to quantum mechanics. To my great surprise I found that black holes seemed to emit particles in a steady state. I put a lot of effort into trying to get rid of this embarrassing effect. What finally convinced me it was a real process was the outgoing [radiation] had a spectra that was precisely thermal."

Hawking had already made several significant contributions to black hole physics but with this discovery he was suddenly thrust into the scientific limelight. Born in 1942, he grew up in London and St. Albans, about 20 miles outside of London. His father was a doctor who did research on tropical diseases. Stephen decided early that he wanted to be a scientist, preferably a physical scientist because he was never particularly drawn to biology despite his father's profession. He said later on, though, that he might have chosen it if some of the recent advances in molecular biology had occurred a little earlier.

Despite his enthusiasm for science he was an indifferent student, rarely taking notes and occasionally falling asleep in class. There is little indication that he was outstanding at school. Upon graduation from high school he applied for admission to study physics and mathematics at Oxford. He easily got through the entrance exam in physics but had a little trouble in the mathematics section. Nevertheless, he was accepted, but his laziness had not decreased and he frequently skipped lectures, referring to them as irrelevant. He did, however, study hard under his tutors, and later said that it was under their guidance that he learned most of what he did.

Upon graduation he decided he wanted to go to Cambridge to work under the well-known cosmologist Fred Hoyle. But when he got to Cambridge he was assigned to Dennis Sciama. Within a short time after starting graduate school, Hawking began noticing that his speech was becoming slurred and he was having difficulty walking. The problem was soon diagnosed as the neural disease amyotrophic lateral sclerosis. It progressed rapidly at first and Hawking became depressed. His studies slipped as he questioned the purpose of studying so hard; he was sure he would die before he got his Ph.D. But gradually the rapid deterioration slackened; this, combined with his marriage to Jane Wilde, gave him hope, so he buried himself in the scientific problems of the day.

Today, although he is confined to a wheelchair, and speaks with difficulty (only those close to him can understand what he is saying), he is still working on some of the foremost problems of physics. He has made several major breakthroughs and has been awarded four honorary doctorates. Some scientists have even compared his contributions to those of Einstein. All of this was accomplished despite the fact that he could not move his arms. He has an automatic page turner, but someone must put the book in it. Most of the time, though, rather than use the turner he prefers to have somebody photocopy the pages of a journal or book he needs and spread them out on a table before him.

Because he cannot use a pencil to make his calculations on paper, he must do them in his head. He therefore has to rely heavily on his memory. But he admits he prefers to translate the problem into geometric diagrams, as much as possible, rather than try to work with mathematical equations. He leaves much of the detailed calculations to his collaborators.

What Hawking showed in relation to black holes was that their surface temperature arises from a strange type of particle evaporation that occurs just outside their surface. This was difficult for scientists to accept because the black hole itself was not composed of particles of any type—all its matter had been crushed in its singularity. How could it create particles?

To answer this we must begin by considering the vacuum. We usually think of it as completely void of particles, but scientists have known for years that this is not the case; space is actually a beehive of activity; particles are being continuously generated in large numbers. But considerable energy is required to generate particles. Where does it come from? As we saw earlier, there is a certain fuzziness associated with nature on the smallest scale, which also applies to energy. It turns out that according to quantum theory the principle of conservation of energy need not be obeyed if the energy is borrowed and paid back in a sufficiently short time. And this is, in fact, what occurs in space. Pairs of particles (a particle and an antiparticle) are generated briefly, then come together again annihilating one another. This occurs so fast that we cannot observe these particles directly and therefore call them virtual particles.

Suppose, though, that just as they were generated we abruptly separated them; they would then be observable. This can be done using a sufficiently large electric field. We know that an electron is pulled in one direction in a field of this type, and the positron is pulled in the opposite direction. If a pair of this type is generated in a capacitor (two plates of opposite charge with an electric field between them), where the field is large there should be numerous particles of opposite sign produced, and indeed there are. In the case of black holes we have a similar phenomenon separating the particles. Just outside the event horizon are strong tidal forces. When virtual particles are generated within this region, they are immediately forced apart because of these forces and consequently become real. Most of these particles will fall into the black hole, but some will escape, so from a distance it will appear as if the black hole is emitting particles. Since many of the particles and antiparticles will annihilate one another on the way out, there will also be considerable radiation.

But this release of energy must come from somewhere. We now know that it comes from the black hole itself. As mass and energy are emitted, the black hole decreases by the same amount; this means it must get smaller. And Hawking has shown that as it gets smaller it emits radiation and particles at an ever-increasing rate; in effect, it gets hotter and hotter.

If we make a few simple calculations, though, we soon see that this evaporation process is insignificant for large black holes (those a few miles across). For these the surface temperature is only a few millionths of a degree above absolute zero, and consequently they evaporate very slowly. For a black hole of mass

10^{20} grams, however, the surface temperature would be 3 million degrees. Interestingly, this black hole would only be the size of an atom, yet it would be white hot, and would emit (release radiation) almost as if it were a white hole. Indeed, Hawking has shown that a mini black hole would be impossible to distinguish from a mini white hole; they would both be "gushers."

This evaporation would continue at a leisurely rate for these tiny black holes; despite their size they have so much energy it would take billions of years to radiate it all away. Obviously, if we could somehow capture one, it would be an extremely useful energy source. In the last moments of the black hole's life, the rate of emission would be so great that it would constitute an explosion. In fact, those that were formed with a mass of 10^{15} grams should be exploding right now. Although the explosion might be the equivalent of a million-megaton bomb, it is small by astronomical standards and would be difficult for us to detect unless it were relatively close (i.e., within the solar system).

A question that immediately comes to mind when we mention exploding black holes is: What remains after the explosion? The event horizon would of course disappear, but there are indications the singularity at the center would remain, but it would now be naked. Whether or not "naked singularities" actually exist in the universe we are not certain, but if they do they present us with a crisis: as in the case of wormholes they may violate causality.

THE LINK BETWEEN GENERAL RELATIVITY
AND QUANTUM MECHANICS

One of the most important aspects of Hawking's discovery of the emission of particles and radiation from black holes was the way they were emitted: the black hole actually acted as if it were a heated object in equilibrium with its surroundings. Hawking showed that the radiation spectrum from black holes obeys the same formula that Planck derived for the emission of radiation from a heated object. And since Planck's formula is quantum mechanical and the black hole is described by general relativity, we appear to have the first link between these two important theories. We do not yet know what the full significance of this link is but it is compelling and it seems as if it might eventually lead to an understanding of the relationship between the two theories and perhaps their unification.

In summary, we see that black holes are particularly important in relation to our objective: a unified theory. This is where general relativity breaks down and where a possible link with quantum theory occurs. But black holes are important in another respect. In the next chapter we will see that the universe began about 18 billion years ago as an immense explosion. This explosion presumably issued from a singularity—the same type we have within the black hole. There is, as we will see, a strange resemblance between the events of the early universe and the processes that occur in the evaporation of a black hole.

CHAPTER 6

The Early Universe

We just saw that general relativity breaks down at the singularity of the black hole. Now we will see that the early universe may have been a singularity, and general relativity therefore also breaks down here.

We live in an expanding universe, a universe that began its expansion, according to the big bang theory, about 18 billion years ago as an explosion of unimaginable intensity. In the first few moments after the explosion there were no stars, no planets, no galaxies—nothing but particles, radiation, and black holes. In short, the universe was in a state of complete chaos, its energy so high that particles were colliding almost continuously with devastating force. It was, in essence, a gigantic "atom smasher" far superior to anything we have here on earth.

Scientists are currently building larger and larger atom smashers (they prefer to call them particle accelerators) in an effort to see what happens when very energetic particles collide. But large accelerators are expensive, and usually take many years to build. Because of this some scientists have become impatient and have turned to the early universe. It has even been jokingly referred to as the "poor man's accelerator" but this is not really a very appropriate name. If we were to try to build an accelerator that could produce energies equivalent to those in the very early universe, it would extend beyond the nearest stars.

There is obviously no way we are ever going to build an

accelerator this large, so the early universe, or at least our models of it, serve an important purpose in helping explain what happens at these tremendous energies.

But why are the events that occur at these energies so important? First, they help us understand the nature of the fundamental particles, but they also help us understand the fundamental forces of nature. An understanding of the link between these forces is a major part of comprehending cosmic ties, and according to recent theories this link between the forces may reside in an understanding of the early universe. Why, we might ask, are there four fundamental forces?—one would be more natural. While we are at it, we might ask the same question of the fundamental particles.

One fundamental force and one fundamental particle, you might agree, would make the universe much simpler. And, as we shall see, this may be the case. According to recent theories the four fundamental forces may merge into one at energies that were present in the early universe. As the universe expanded and cooled there may have been a separation of forces; just as water freezes out as the temperature lowers, gravity may have "frozen" out of the unified force. This would have left the three remaining ones unified. But shortly thereafter the weak nuclear force would have frozen out, then finally the electromagnetic and strong nuclear forces would have separated. If our idea is correct and there is a unification at high energies, a study of the early universe would be of supreme importance. Before we look into events of the first seconds after the explosion, however, we must consider the evidence we have that there was, indeed, an explosion.

DISCOVERY OF THE EXPANDING UNIVERSE

Strangely, the man who was indirectly responsible for the discovery of the expanding universe died without ever knowing about it. In fact, even if he had lived long enough to be told, he would likely not have cared, for he was not interested in stars or galaxies (which were not known to exist at that time). He was interested in planets—Mars, in particular. His name was Percival Lowell.

Lowell had little interest in astronomy before he was 35. But shortly after seeing some of the first drawings of Mars by the Italian amateur astronomer Giovanni Schiaparelli, his imagination was set into motion. The drawings showed "canals," which implied that there might be intelligent life on the planet. His interest became so intense that he decided in the early 1890s to set up an observatory to study Mars, and he set off to the west looking for the clearest, darkest skies that America offered. He found them at Flagstaff, Arizona, and within a few years he had constructed an observatory on a knoll just outside of town.

Although he had little interest in anything besides the planets, he had some interest in the "fuzzy" objects in the sky— the nebulae. According to a theory Laplace had published many years earlier, these objects might be the beginnings of planetary systems. Laplace had envisioned our system as evolving out of a gas cloud, and these fuzzy objects looked very much like gas clouds. Lowell hired V. M. Slipher to see if they were.

Slipher was in many ways the antithesis of Lowell. Where Lowell's imagination ran wild with flights of fantasy and speculation, Slipher was cautious, methodical, and careful. His first task was to find out if the fuzzy objects were rotating. To do this he had to use the spectroscope, an instrument in which the light from the object is passed through a prism (or a grating) so that the various colors can be separated. When the light from a star or nebula is passed through this instrument, we get a series of lines, the most prominent of which are caused by the hydrogen in the object, but helium, carbon, sodium, and other elements also produce lines. We know where each of these lines normally lies, but if the object is moving relative to us, the lines will be shifted in one direction or the other. If the object is moving away from us there is a shift toward the end of the spectrum where the lines are normally red; we refer to such a shift as a "redshift." If the object is moving toward us there is a blueshift.

If Slipher's nebulae had been rotating he would have got a redshift from one edge (the one receding from us) and a blueshift from the other. But this was not what he got—and what he got surprised him. He began his study with the bright nebula in Andromeda and found that it exhibited only a blueshift, which meant it was approaching us. He continued studying other bright nebulae and found that he obtained blue- and redshifts (but never both in the same nebula)—some of these nebulae were obviously moving away from us, others toward us. But as he extended his study to dimmer nebulae that were farther away, he found that all of them exhibited redshifts.

In 1914 he presented his findings at a meeting of the American Astronomical Society. He was still uncertain of their meaning and was therefore cautious about speculating on them, but his slides showed the results clearly. The crowd must have realized that something significant had been discovered for when he finished he was given resounding applause. A number of astronomers soon verified his discovery but strangely nobody took up the study seriously, and for another 10 years Slipher had the field to himself.

It might seem that Slipher should have realized the significance of his discovery, but in retrospect we must remember that astronomers were still not certain what nebulae were. Some, like Lowell, thought they were gas clouds, perhaps solar systems in formation, while others thought they were island universes of millions of stars.

Slipher did have some idea of the consequences of his work, though, for in 1921 he wrote in the *New York Times:*

> . . . The lines in its spectra are greatly shifted, showing that the nebula is flying away from our region of space with a marvelous velocity of 1,100 miles per second.
>
> This nebula belongs to the spiral family, which includes the great majority of nebulae. They are the most distant of all celestial bodies, and must be enormously large.
>
> If the above swiftly moving nebula is assumed to have left the region of the sun at the beginning of the earth, it is

easily computed, assuming the geologist's recent estimate of the earth's age, that the nebula must be many millions of light years distant.

The velocity of this nebula suggests a further increase in the estimated size of the spiral nebulae themselves as well as their distance, and also swells the dimensions of the known universe.

It is perhaps ironic that Slipher was within grasp of one of man's greatest discoveries, yet never made it. The idea that the universe was expanding apparently never occurred to him even though by 1923 he had measured the spectral shifts of 45 nebulae, and almost all of them were redshifted. By about 1925 he had come to the limit of his rather meager 24-inch telescope and could go no further; he therefore moved on to other projects.

The man to whom the torch was passed was in the audience when Slipher first announced his discovery in 1914, but at the time he was still a graduate student. The discovery must have had a particular significance to him even at that time, since his thesis project was on nebulae—the very objects Slipher was talking about. His name was Edwin Hubble.

Hubble is in all respects the astronomer's astronomer. Born in Marshfield, Massachusetts, in 1889, his abilities in the classroom and on the playing field were evident from an early age. He was both an outstanding scholar and athlete even in high school, and when he went to the University of Chicago to study physics he never slowed down. It was said that he learned almost too easily, and seemed to excel in the classroom without studying. This, of course, left considerable time for athletics and he was outstanding in most: track, basketball, boxing, and rowing. His undergraduate abilities were, in fact, outstanding enough to win him a Rhodes scholarship, which he used to study law at Oxford.

On his return to the United States in 1913 he set up a law practice in Louisville, Kentucky, but within months he became bored and dissatisfied with law. He had had an introduction to astronomy in his undergraduate days at the University of Chicago, and as a youth had learned the constellations and had read popular astronomy books. The yearning to get out of law became greater as the months passed and finally he made a decision: Even if he were to become a second- or third-rate astronomer, it was astronomy that really mattered to him. He decided to chuck law and return to the University of Chicago. At Chicago he signed up for the graduate program in astronomy and was soon working with the giant telescope of the Yerkes Observatory. Nebulae were his first objects of study, and he eventually devoted his life to them.

He received his Ph.D. in 1917 after staying up all night finishing his thesis and cramming for his oral. An offer had already come from Mt. Wilson Observatory and he could have been on his way to a new job within days, but there was a problem—the War. Instead of going to Mt. Wilson, he enlisted and was soon shipped to Europe. He was commissioned captain upon entering but rose rapidly to the rank of major. A wound near the end of the war took him out of action and he returned to the United States about a year after Armistice Day.

Upon his return he went to Mt. Wilson to begin what would become the most extensive study of nebulae that had ever been undertaken. Slipher's discovery was no doubt on his mind when he began but there were other problems to be dealt with first. Astronomers still did not know what nebulae were. His studies at Yerkes had convinced him that they were island universes of stars but he needed proof. He began by taking long-exposure photographs of some of the nearby nebulae, including the Andromeda nebula, and eventually his persistence paid off. The arms of some of them were finally resolved into stars, proving they were not patches of gas; the island universe hypothesis was correct. But there was still the problem of their distance: How far away were they? To be distant systems they had to be well outside our system, the Milky Way. Hubble studied the resolved stars and noticed that some of them were Cepheid variables—stars that changed periodically in brightness. This was the breakthrough he needed, for several years earlier Henrietta Leavitt and Harlow Shapley had established that there

Edwin Hubble working with the 100-inch telescope at Mt. Wilson. (Courtesy Huntington Library, San Marino, California.)

was a relationship between the period of a Cepheid and its distance: If we knew its period (time between equal brightness) we could determine its distance. Hubble made the calculation and showed that the nebulae were indeed well outside the Milky Way—they were galaxies, separate and distinct from ours. Although a few astronomers refused to accept Hubble's results immediately, within a short time the nebula debate was settled once and for all.

With the question of their nature out of the way, Hubble could turn to Slipher's results. What was the significance of the redshift he had discovered? Did it apply to all galaxies, and if so, what did it mean? Along with his assistant, Milton Humason, he set out first to verify Slipher's results, then to extend them. It was a task that required many years, but Hubble persevered, carefully photographing the spectra of dimmer and dimmer galaxies. Soon after beginning the work, however, Hubble realized that an important key was missing: he did not know the distance to most of the galaxies he was photographing. In a few of the nearby ones he could use Cepheids, but the 100-inch telescope could not resolve even moderately distant ones. As he focused his attention on this problem, he turned more and more of the work of obtaining the spectra over to Milton Humason.

Humason's jovial, warm personality contrasted sharply with Hubble's. Where Hubble was considered by many to be distant and unfriendly, Humason was liked by everybody. He appeared at Mt. Wilson one day in search of a job, but because of his lack of training he was offered the lowly task of a mule packer. In those days the only way up to the Mt. Wilson Observatory was via a rough, steep, and winding road, a road only mules could easily negotiate. So supplies came up by mule trains. Humason led one of these mule trains for a few months, then graduated to the position of janitor of the observatory. But he was not going to let his ignorance hold him back; every chance he got he helped the astronomers at the telescope, and while he helped he bombarded them with questions. It soon became evident that his abilities far exceeded those of most janitors and he was taken on as an observer's assistant. Though he had virtually no training in astronomy and had to learn everything on his own, he developed within a short time into a skilled observer.

As Hubble pondered the problem of distance and poured over the results, Humason struggled to get spectra of increasingly dimmer galaxies. With persistence and ingenuity Hubble developed a "cosmic ladder." Knowing the distance to the nearest galaxies—ones with Cepheids in them—he could use them as a stepping-stone to more distant galaxies. The Cepheids were what he called his "primary indicators"; as "secondary indicators" he selected the brightest star in galaxies and assumed they were all about the same brightness. This allowed him to determine their distances and to step out to even more distant galaxies. Finally, as "tertiary indicators" he used the brightest galaxies in groups of galaxies, again assuming they were all about the same brightness. Once he had calibrated his cosmic ladder, he had a method for reaching exceedingly dim galaxies. But there was a problem: each rung of the ladder depended on the rung below it. If there was an error in one, it propagated to the one above it and threw his calculations off. Nevertheless, the method seemed reasonable (several corrections were made in later years) and after several years of hard work Hubble finally announced his results in 1929: the universe was expanding. The galaxies themselves were not changing, but the space between them was increasing linearly with time. This meant that galaxies were moving away from us, and the farther the galaxy was away the faster it moved. His original announcement was based on results that were somewhat ambiguous and many people felt that he was influenced by the theoretical work that was being done in Europe—it predicted that the universe should be expanding. But by 1931 there was no doubt; his observations showed a clear relationship between distance and velocity.

The theoretical work in Europe had begun many years earlier but scientific news traveled slowly in those days and there was little contact between observational astronomers and theoreticians working on the same problem. Einstein began working on cosmology shortly after he finished his general theory of relativity in 1916, but in applying his theory to the entire universe he ran into difficulties. Astronomers had assured him that although there was considerable random motion of objects in the universe, on a large scale it was static, and Einstein considered this to be prerequisite to his theory. When he tried to solve the resulting equations, though, he found that his universe either expanded or contracted. It therefore had to be held fixed, and the only way he could do this was by introducing a constant into his equations. He was reluctant at first, feeling that it would destroy their simplicity and beauty. But eventually he relented and satisfied himself that on a universal scale things were different. Although his "cosmological constant" was important for the universe as a whole, it was negligible when the theory was applied to ordinary astronomical objects.

The universe that Einstein constructed was spherical; a light ray moving in a straight line in a particular direction would eventually trace out a gigantic circle and arrive back at the point from which it was emitted. This property resolved a problem that had plagued astronomers for years: Where is the edge of the universe? And, indeed, if it has an edge, what is on the other side? In Einstein's universe we did not have to worry about an edge—it has no edge, yet it is closed. This meant that it was not infinite in extent—a relief to many.

In the same year (1917) that Einstein published his cosmology, another one was published by the Dutch astronomer Willem de Sitter. Einstein had sent de Sitter a copy of his article on general relativity the year before, which had so fascinated de Sitter that he had passed it on to Eddington in England. Eddington was equally fascinated and later did much to publicize the theory.

De Sitter was born in the Netherlands in 1872. Upon graduation from high school he went to the University of Groningen intent on studying mathematics but was soon drawn to astronomy. After receiving his Ph.D. in 1897, he spent 2 years in Capetown, South Africa, observing southern skies. A few years after his return to Europe, he obtained a professorship at the University of Leiden, and in 1919 he was named director of the observatory.

De Sitter's model was, to say the least, strange. Its strangest property was that it was empty. "But after all," he frequently said, "the real universe is almost empty." Another strange property was that it predicted a redshift (this is what may have influenced Hubble). De Sitter's model got considerable interest for several years—perhaps more than it deserved, considering its properties. Although Einstein became a close friend of de Sitter he never liked his theory. An empty universe somehow never appealed to him, and its prediction of a redshift confused and made little sense to him.

De Sitter had retained the cosmological constant in his theory because he, like Einstein, believed that it was required for a static universe. We will see later, however, that de Sitter's model was not static. Not everyone was convinced, though, that the cosmological constant was needed. At some distance away, in the USSR, isolated from the other relativists, was the mathematician Aleksandr Friedmann. Most of his career had been in applied areas: he was a mathematical assistant at the Institute of Bridges and Roads and later lectured at the Institute of Mines. But his work in hydrodynamics involved the mathematical techniques of tensor analysis, and as a consequence he was drawn to Einstein's theory, which was written in the language of tensors.

Upon examining Einstein's cosmology (without the cosmological constant), Friedmann found that one of the large expressions in the denominator went to zero under certain circumstances, something that was apparently overlooked by Einstein. He followed the details through and discovered an interesting evolutionary (time dependent) theory of the universe. Excited about his results he sent them to Einstein but received no reply. After several months he decided to go ahead and publish them. His paper, which appeared in the German journal *Zeitschrift für Physik* in 1922, soon came to Einstein's attention and he wrote a short note to the editor criticizing it (he believed there was an error in it). The editor published the note in the next edition, but Friedmann soon saw that Einstein's criticism was unfounded and pointed it out. Einstein blushingly retracted his criticism with another note, but for some reason he never wholeheartedly supported the theory.

Friedmann's paper got little attention despite being published in a prestigious journal. There are perhaps two reasons for this. First, Slipher's results were not yet known in Europe and, of course, Hubble had not even started working on the problem. In essence, there was no reason to believe in an evolutionary universe. Second, Einstein's criticism may have had some effect. Einstein no doubt disliked the result because he was convinced the universe was static and an evolutionary model was of little value; after all, he had introduced the cosmological constant to get around that type of model. It is perhaps strange, though, that after it became clear the universe was expanding and the solution was valid that Einstein did not immediately bring it to people's attention. Perhaps the fact that he had missed the solution may have perturbed him, or maybe he just forgot about it.

Friedmann published a second paper on the subject 2 years later, then died the following year (1925) of typhus fever. He never lived to see the results of his labor. Although his theory was literally buried for several years, it finally came to the attention of the scientific world and is now the theory that is in general use.

Within Friedmann's model there are three possible universes, each curved in a different way. The first is a positively curved universe (Riemannian geometry), much like Einstein's, that expands to a certain radius and then collapses back on itself. The second universe is a negatively curved one (Lobachevskian geometry) that expands forever, and between these two is one based on Euclid's geometry, a flat universe. It also expands forever. Which of these models is correct depends on the average density of matter in the universe. If it is over a certain critical value, the universe is positively curved and will eventually collapse back on itself; if it is under this value, it is negatively curved and will expand forever. Unfortunately, we are still uncertain of the average density and therefore do not know the future of our universe.

At the time Friedmann's papers were published, de Sitter's universe was still getting considerable attention throughout Eu-

rope. A few years later Hermann Weyl showed that if two particles were placed in it they would fly apart, and the farther they
separated the faster they moved. Since this applied to galaxies,
it was obvious that de Sitter's universe was not a static one after
all. It was, in fact, used to predict the expansion of the universe,
even before the expansion was discovered by Hubble.

But de Sitter's universe was an empty one, a point that
disturbed many, including a Belgian priest named Georges
Lemaitre who had just begun working in cosmology. Lemaitre
looked more closely at Einstein's equations and discovered another evolutionary model. It was different from the previous
evolutionary models in that it had many possible models within
it. Lemaitre selected from them one that particularly appealed to
him: it was, in a sense, a combination of Einstein's and de Sitter's models. It began in a state of explosion and expansion,
then slowed down and became stable for a while (Einstein case).
Lemaitre believed that galaxies may have formed during this
stable period. Eventually, though, it became unstable and began
expanding as the de Sitter model does. Lemaitre's paper was,
unfortunately, published in a rather obscure journal and it drew
little attention.

Hubble's announcement of 1929 excited the astronomical
world. If the universe was expanding, theories should predict
it—and of course de Sitter's did, but Eddington was not satisfied with it. He published a note about the need for a good
evolutionary theory that would explain the results. Lemaitre
saw it and got in touch with Eddington immediately, telling him
that he had published such a theory a few years earlier. Eddington read Lemaitre's paper with enthusiasm, and was so
delighted with it he had it republished in *Monthly Notices*. About
this time he began looking into the problem for himself and
soon discovered that even Einstein's model was not truly static.
It was in a state of unstable equilibrium: a slight jar in one
direction and it would begin to expand, or a jar in the other
direction and it would contract.

Within a short period of time Friedmann's papers also came
to Eddington's attention and it was soon established that his

theory (without the cosmological constant) was the most satisfactory. Even Einstein eventually cursed himself for ever introducing the cosmological constant, calling it the "worst blunder
of his life." Today most astronomers use Friedmann's theory in
a slightly modified form, a form that was introduced independently in 1935 by two American physicists, Howard P. Robert-
son and Arthur Walker.

Let us return now for a moment to Hubble's work and look
more closely at what he accomplished. We saw earlier that he
developed a cosmic ladder to the outer galaxies that allowed him
to determine their approximate distances. Once he knew both
their distance and speed, he made a plot of these two quantities:

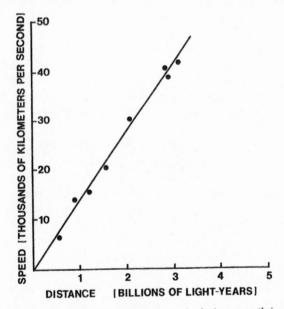

The Hubble diagram—a plot of the distance of galaxies versus their speed.

The points showed a fair amount of scatter but a straight line
could be drawn through them giving a relationship between

speed and distance. The proportionality factor between these two quantities (which is equal to the slope of the line through the points) is now known as Hubble's constant. Over the years, as corrections and technical advances have been made, the accepted value for this constant has changed considerably.

Hubble summarized his results in a book titled *Realm of the Nebulae* in 1936, a book that is now considered to be a classic. Along with Humason he pushed the 100-inch telescope to its limit. Then with the completion of the 200-inch telescope at Palomar in 1948 he continued studying even dimmer galaxies. In 1953, he died before his work was complete.

By the early 1930s the expanding universe model was accepted by the majority of cosmologists. Galaxies, or at least groups of galaxies, were all expanding away from us—the farther they were away, the faster they traveled. (Galaxies within groups do not expand away from one another because their mutual attraction is greater than their universal repulsion.) Since all galaxies are traveling away from us, it might seem that we are at the center of the universe, but this is not the case. It is the space between the galaxies that is expanding, so regardless of where you are in the universe all galaxies will appear to be moving away from you.

But if the universe is presently expanding, it is easy to see that it must have had a beginning. And this means that if we reverse time (let it run backwards) the universe will contract; in fact, it will continue contracting until all the matter of the universe is in one place—a seemingly strange result that neither Einstein nor Eddington liked. Eddington preferred to believe that the universe was originally in an Einstein state, i.e., static, when suddenly something disrupted it and it began to expand. This gets around both the problem of a beginning and a dense primordial state. But Lemaitre, perhaps because he was a priest, and the Church preferred a beginning to the universe, was fascinated by the possibility of a dense origin. He referred to this original condensed universe as the "primeval atom." George Gamow, who later extended Lemaitre's ideas, remarked in his book *The Creation of the Universe* that Lemaitre should have called it the primeval nucleus. Actually, Lemaitre did think of it as a gigantic nucleus rather than as an atom, believing that it fragmented, or fissioned just as uranium fissions in an atomic bomb. This fragmenting continued until the universe was filled with elementary particles. He described the process in his book *The Primeval Atom* (1951):

> The atom would break up into fragments, each fragment into smaller pieces. Assuming, for sake of simplicity, that this fragmentation occurred in equal pieces, we find that two hundred and sixty successive fragmentations were needed in order to reach the present pulverization of matter into our poor little atoms which are almost too small to be broken down further. The evolution of the world can be compared to a display of fireworks that has just ended: some few red wisps, ashes and smoke, standing on a cooled cinder, we see the slow fading of the sun, and we try to recall the vanished brilliance of the origin of the worlds.

Lemaitre worked on his ideas for several years. When the theory was later referred to as the big bang theory by Fred Hoyle (and popularized as the big bang by Gamow), Lemaitre became known as the father of the big bang. Detailed mathematical treatment of Lemaitre's ideas showed, however, that there were difficulties and because of this a new approach soon came into being.

George Gamow was the initiator of this new approach, an approach that is quite close to what we believe today. Born in Russia in 1904, Gamow was already absorbing Jules Verne's books and dreaming of flights to the moon by the age of 7. Despite a rather sporadic elementary education—school was frequently closed because of the war—he developed an intense interest in astronomy and physics. By the time he was ready to go to college, the war was over and things had calmed down, but the aftereffects remained. He went to Novorossia University in Odessa hoping to major in physics and mathematics, but the physics department had folded because the only professor available refused to teach under the deplorable conditions. Gamow therefore concentrated on mathematics, but even here there were problems: most of the lectures were in the evenings and the lights frequently went out. But as Gamow says, "the professors would just keep on lecturing."

After a year of this he had had enough and decided to leave for the University of Leningrad. At Leningrad he became fascinated with Einstein's theories, and though there was a professor there who was familiar with them, he died shortly after Gamow arrived.

In 1928 Gamow left Leningrad for the University of Göttingen in Germany, the European center of theoretical physics. It was an exciting time as quantum mechanics had just been developed and some of the best physicists in Europe were at Göttingen. Gamow got caught up in the activity and published one of his most important works: the use of quantum mechanics to explain α decay and quantum tunneling.

George Gamow (1904–   ) (to the left). (Courtesy AIP Niels Bohr Library, Uhlenbeck Collection.)

While he was at Göttingen he visited Niels Bohr at Copenhagen; he planned on staying only a short time, but Bohr invited him to stay a year, which he did. Then it was on to Cambridge for another year or so, and finally back to Russia. Soon after arriving in Russia, though, he decided it had been a mistake to come back, and within a short time he was concocting plans to escape. There was a slight complication now, though: he had recently married.

His first plan was to cross the Black Sea to Turkey in a small kayak. Outfitting it with 5 or 6 days of food they set off, Gamow in the front and his wife in the back. The first day was smooth sailing but by the time they got out of sight of land the wind began to blow and waves started to develop. Pretty soon the waves were washing across the small kayak, putting them in serious danger of sinking. While Gamow paddled his wife bailed. Finally, both exhausted, they fell asleep and when they woke the storm was over—but there was still no sight of land. Gamow had had enough by now; he turned the kayak in the direction he thought was closest to land, and paddled for all he was worth. They soon sighted the coastline but upon landing discovered they were still in Russia.

Somewhat later a second escape plan was concocted: they were going to ski across a rugged snowy section of Russia to Finland. But that plan never got off the ground. Then, much to his surprise, Gamow got a letter from the government ordering him to attend the Solvay conference in Brussels as a representative of Russia. He jumped with joy, almost unable to believe his luck.

In Europe, Gamow traveled the countryside on his motorcycle generally enjoying himself before finally coming to George Washington University in the United States. Gamow is famous both as a prolific popularizer and for his irrepressible sense of humor; he was known to quickly take advantage of any opportunity to play a practical joke.

Gamow was attracted to the ideas of the early universe through his interest in the origin of the elements. How were the various elements in the universe formed? Earlier it was believed that they were produced in stars, but in 1939 Hans Bethe shook up scientists by showing that only elements up to helium could be produced in this way (a conclusion that was later shown to be incorrect). A suggestion was made in 1942 by Chandrasekhar that they may have been produced in the early universe; because of the high density at that time, the temperatures should have been greater than 10 billion degrees—easily high enough to produce nuclei.

Gamow followed up on Chandrasekhar's suggestion, but he took a different approach than Lemaitre. Where Lemaitre had assumed a fissioning (or fractionation) of the primeval nucleus, Gamow assumed a nucleus in which fusion (or joining together) of particles occurred, as they do in the hydrogen bomb. According to Gamow: "The original state of matter is assumed to be a hot nuclear gas (not a fluid). It is also assumed that physical conditions at that epoch were changing so rapidly that no real equilibrium was ever established. . . ." His nucleus was a mixture of neutrons, protons, and electrons at exceedingly high temperatures. It is well known that free neutrons decay in about 13 minutes into protons and electrons, but temperatures would have been so high that when an electron slammed into a proton it would have created a neutron, so that there would have been, according to Gamow, a kind of quasi-equilibrium state. Gamow referred to this chaotic mixture by the colorful but little-used noun, *ylem* (which means first substance, from which elements are formed). He realized that originally the temperatures would have been too high for nuclei to form from these particles, but eventually as they cooled below $10^9\,°K$ nuclear reactions could occur. The time during which they would occur, though, would be relatively short—probably no longer than an hour—because of the expansion and subsequent cooling of the universe. The temperature would eventually be too low for the creation of neutrons via proton–electron collisions and all neutrons would soon disappear from the universe.

About this time nuclear physicists had begun to work out

the details of nuclear reactions of the type occurring in the early universe and the probabilities of occurrence (cross sections) of some of them had been calculated. All Gamow therefore needed was a good graduate student to carry out the tedious calculations (the dirty work always goes to the student)—and one soon became available. Ralph Alpher had just been scooped by the Russian physicist Lifshitz in a thesis project related to galaxies and was in need of a new one. Gamow therefore assigned him the task of investigating how, starting with the ylem, the various elements could have been built up by consecutive bombardment by neutrons. Alpher took the data points he had available (cross sections), drew a smooth curve through them, and proceeded. Soon he had shown that the elements could be built up in a stepwise process as Gamow had suggested.

When they went to publish the paper, Gamow—always on the alert for a pun—noticed that his and Alpher's names sounded like the first and third letters of the Greek alphabet: alpha and gamma. To complete the sequence he needed a beta, and indeed an old friend at Cornell had the name he needed: Hans Bethe. So Gamow added it to the list, and the theory eventually became known as the αβγ (alpha–beta–gamma) theory. Gamow says that Bethe did not mind at all and later was quite helpful in discussions of the theory, but when it was eventually found to be at fault Gamow said (with a smile) that he heard a rumor that Bethe was planning on changing his name. Speaking of name changes, Gamow also claims he asked Herman, who later worked on the project, if he would change his name to Delter so he would have a more complete sequence (delta being the fourth letter of the Greek alphabet), but he said Herman "stubbornly refused." So much for Gamow's puns.

Shortly after the publication of the αβγ theory it came to the attention of Enrico Fermi. Fermi was unhappy with the way Alpher had drawn a smooth curve through the data points. Using a better set of points that were available to him (and not smooth, particularly in the case of light elements) he had a student (A. Turkevich) redo the problem in detail. Turkevich found that Gamow's scheme worked only to helium; at helium there was a gap that could not be hurdled (there was also another at slightly heavier elements). Alpher and Gamow noticed this about the same time. Thus, it appeared as if the heavier elements could not be generated in the early universe nor in stars. Bethe had earlier found a similar problem in stars.

But much more was now known about the reactions in stars so at the suggestion of Fermi, Martin Schwarzschild began looking at the spectra of stars to see if there was any evidence for the generation of heavy elements. And indeed he found some. The problem then was to explain how the gap was mysteriously hurdled. He assigned the task to his student Edwin Salpeter in 1951, and Salpeter soon showed that there was a way: a series of reactions using beryllium (which would be available in stars) would allow the generation of carbon from helium.

One of the important predictions of Gamow's theory was the temperature of the universe. Radiation from the big bang explosion had spread out into the universe and cooled, but it still had a temperature of about 25°K according to Gamow. Alpher and Herman later redid the calculation and determined that this temperature was only about 5°K. They believed, however, that with the technology of 1948 it was impossible to detect such a low level of radiation and therefore never searched for it, or encouraged anyone else to search for it. They were sure, in fact, that it would be camouflaged by the energy associated with starlight.

Gamow's theory faded away rapidly in the early 1950s when scientists discovered that the elements could be made in stars, but about a decade later there was a resurgence of interest in it. Fred Hoyle was examining the helium content of the universe when he made a remarkable discovery: it was impossible that all of the helium of the universe was made in stars; most of it—up to 90%—had to be made elsewhere. The early universe was the obvious candidate, and within a short time it was shown that most of the helium was produced there.

By the mid-1960s it was generally accepted by astronomers

that the universe did begin as a big bang, and that at one time it was infinitesimally small. Most people find it extremely difficult to accept the fact that the entire mass of the universe was at one time contained within a nucleus that is smaller than an atom. But there is something else about this primeval nucleus that is even more difficult to comprehend. We would tend to believe that it existed in some sort of infinite space and exploded in it, but this, according to astronomers, is not the case. There was no space around this nucleus: the nucleus was the universe. When it exploded it created space, time, and matter. Later we will trace the details of this explosion and see how the universe evolved from it, but first let us go back in time to this explosion.

BACK TO THE BIG BANG

To go back in time to this first instant we must know the age of the universe. Unfortunately, there are still problems in relation to this number. For now let us just take the most generally accepted number: 18 billion years. This means that 18 billion years ago our universe began as a gigantic explosion—the big bang.

Today galaxies are receding from us, and with a reversal they would begin moving toward us: the universe would, in effect, begin to shrink. Galaxies are so widely separated now, though, that it would take about 16 billion years before they got very close together. Assume, then, that we, as immortal beings, are traveling back in time, with billions of years passing in minutes. We see lights—stars—blinking off and on in our galaxy, new stars forming from the gas and dust of space, living out their lives and perhaps exploding and dissipating off into space, or just fading slowly away. From a distance it might look like blinking Christmas tree lights. As we continue back in time the intensity of the lights in some of the galaxies increases slightly, but gradually each galaxy has more and more gas and fewer stars. Finally the last stars blink out and there is nothing left but a gigantic swirling mass of gas. Each of the giant gaseous spirals increases in size as it gradually approaches other spirals, then when the universe is a few hundred million years old the giant blobs dissipate, and all of space is filled with an exceedingly thin, but generally uniform gas. There are, however, small but perceptible fluctuations within it; astronomers are not yet certain exactly what caused the breakup of this space-filling gas, but it seems that some sort of shock wave generated within a few seconds (or perhaps minutes) of the explosion was responsible.

When the universe was about 10 million years old it was at what we now refer to as room temperature. Though it seemed to be empty, the sky was the deepest black imaginable, without a single visible object in it. But there was something there—the thin, diffuse matter of galaxies.

As we move back further in time the gas continues to heat, and within a few million years there is a slight glow, which gradually turns deep red; the temperature at this stage is about 1000°K. The universe is eerie yet transparent and uniform; gradually, though, it changes to orange, then yellow. Then suddenly at a temperature of 3000°K something strange happens. Up to this point the universe has been transparent—there was nothing in it to see, yet we could still see through the dim colored mist. But now a brilliant yellow fog suddenly envelops us, cutting off our view—we cannot see a thing.

Continuing back even further in time we find that the universe is composed almost entirely of dense radiation, yet a few nuclei are buried in the mist. The brightness of the mist intensifies as the heat continues to increase. Light particles and their antiparticles begin to appear everywhere—the universe at this stage is a mixture of radiation, electrons, neutrons, and their antiparticles. Finally, at even higher temperatures heavy particles and their antiparticles appear along with black holes. The universe becomes an unimaginable chaotic soup—particles and radiation slamming into one another with incredible force. It is now tiny, perhaps the size of a beachball, and within a fraction of a second it may become a singularity. But before it does, a "screen" descends in front of us. We thus cannot predict what actually happens in this final fraction of a second because we have no way of looking at it. The screen I am referring to is a screen of ignorance: general relativity breaks down and even perhaps quantum theory, and we have no way of seeing what is beyond. We thus cannot say for certain that a singularity does occur.

THE ULTIMATE SINGULARITY

The universal singularity, or near singularity discussed above is similar to that within a black hole. In the type of black hole discussed earlier, however, the singularity had a total mass equal to that of a large star; in this case we are talking about a singularity that contains all the mass of the universe. There is a basic difference aside from this, though. In the case of a collapsed star we have an event horizon with a singularity at the center—a black hole sitting somewhere in our universe. In the case of the universal black hole there is a problem: if our entire universe collapsed to a black hole, all its matter and space would disappear into the singularity. This means there would be nothing left for it to be sitting around in—no universe would exist. Also, in the case of the universal black hole, or perhaps I should call it a "quasi black hole," we cannot be sure we are even dealing with a true singularity.

But even if a singularity did not occur, there is the question of what happened before—much before it. One possibility is that there was another universe that collapsed; it may have collapsed to a singularity or near singularity and then emerged giving us our universe. Indeed, there may have been many such collapses and reemergences. This situation is referred to as the oscillating model of the universe.

Getting back to where general relativity breaks down, we find it happens 10^{-43} second (one divided by one followed by 42 zeros) after time zero, a time interval we usually refer to as

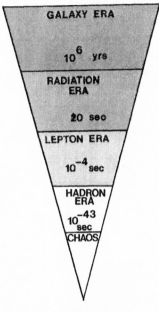

GALAXY ERA

10^6 yrs

RADIATION ERA

20 sec

LEPTON ERA

10^{-4} sec

HADRON ERA

10^{-43} sec

CHAOS

BIG BANG

A simple representation of the eras of the universe beginning with the big bang.

the Planck time. This is, in effect, the point where the curtain is drawn. Immediately after this time the universe is in complete chaos but with the help of quantum theory we can arrive at a crude picture of what probably happened.

Earlier we mentioned Stephen Hawking's contention that tiny black holes were generated in the very early universe; he also proved that these tiny black holes evaporated in about 10^{-43} second. This means that shortly after 10^{-43} second a strange "foam" of black holes existed. David Schramm of the University of Chicago describes it as follows: ". . . we are led to a picture of spacetime as a foam of mini black holes popping . . . , recombining and forming up again." Space and time at this instant were quite different from the way we now perceive them: they were both disconnected. The foam was, in effect, a mixture of space, time, black holes, and "nothing" in a strange disconnected state—a state we know little about.

The temperature at this time (10^{-43} second) was approximately $10^{32}\,°K$—easily high enough to generate particles. There are, in fact, two ways that they could have been generated. In the first method an electron is generated along with an antielectron (called pair production) when there is enough energy available, or equivalently if the temperature is high enough. If the temperature is 6 billion degrees, for example, two colliding photons can produce an electron–antielectron pair. At even higher temperatures, proton–antiproton pairs and so on can be produced: the heavier the particle the more energy required or equivalently the higher the temperature needed.

We saw earlier that there is also a second way particles are generated, again in pairs: they are produced just outside the event horizon of mini black holes as a result of tidal forces. We also saw that as mini black holes evaporated they created showers of particles, and since the universal black hole is similar to this mini black hole, it would create particles in the same way. Thus, we have two methods of generating particles. Which, then, is the more important? According to astronomers, during most of this time interval the generation of pairs via high energies is the more important; only during the very earliest part of it are the tidal forces sufficient to generate large numbers of particle pairs. We still have much to learn about this area, however, and the latter method may eventually be shown to be significant.

The era (brief period of time) immediately after 10^{-43} second is usually called the quantum era. It was during this time that the four fundamental forces of nature were unified. Shortly after 10^{-43} second the unified field fractured and the first of the four broke away. Later, one by one each of the others broke away and changed in strength, giving us finally the four forces we have today.

INFLATION

One of the difficulties of the standard big bang theory is the incredible amount of energy needed for the generation of particles. Where did it come from? A modification of the big bang theory that appears to answer this question has recently attracted a considerable amount of attention. It is called the inflation theory and was put forward in 1980 by Alan Guth of MIT. The major difference between inflation theory and the standard big bang theory occurs from about 10^{-35} second to 10^{-32} second after the explosion. According to Guth, at about 10^{-35} second the universe settled into what is called a "false vacuum," a state in which the energy of the universe was exceedingly large. Because of this false vacuum an incredibly fast expansion— much faster than that of the big bang—occurred. (This is where the name comes from.) The universe at 10^{-35} second consisted of nothing but mini black holes and disconnected space, so when the sudden inflation occurred it pushed not one universe into existence but many, some perhaps contained in others. Each of the separate pieces of the foam became an individual universe and we live in one of them. This means that there may be many universes out there that we can never observe.

Although the theory gets around several of the problems of the standard big bang theory, scientists find it has problems of its own. It is, for example, difficult to stop the inflation once it gets going. This was overcome by a new version of the theory that appeared in late 1981 but even this theory has problems.

THE HADRON ERA

At 10^{-23} second the universe entered the hadron or heavy particle era. Since hadrons interact via the strong nuclear interactions, we can also think of this as the strong interaction era. The temperature was high enough to generate hadron pairs: mesons, protons, neutrons, and so on, and their antiparticles.

At the beginning of this era, though, the temperature was too high for these particles to exist in their usual form, therefore only their components, namely quarks, existed. The universe, at that stage, thus consisted almost entirely of quarks and anti-quarks. At the present time we do not see free quarks; they are theorized to be trapped in "bags" from which they cannot escape. Some scientists believe, however, that there should still be a few residual ones left from this early era. They may, in fact, be as common as gold atoms, but so far we have not found them.

According to this theory when the temperature had decreased sufficiently (at about 10^{-6} second) the quarks suddenly assembled themselves into "bags." We refer to this event as the quark–hadron transition. The universe then consisted mostly of mesons, neutrons, protons, their antiparticles, and photons; there may also have been some heavier particles present along with some black holes. For each particle, though, there was its antiparticle, and whenever a given particle met its antiparticle they would annihilate one another with the release of one or more photons. The photons, however, might then have created new pairs; thus, the universe was in an equilibrium state in which pairs were being created and annihilated at about the same rate. But as it continued to expand it cooled until finally the temperature was too low for creation of pairs. Gradually the annihilations outnumbered the creations until finally almost all the heavy particles were gone. If there had been an equal number of particles and antiparticles present, they would have all disappeared. But we know this was not the case: a small residual had to survive—or we would not be here.

The temperature was finally so low that heavy particle pairs could no longer be generated, but there was still enough energy for the generation of light particles (leptons). The universe then entered an era in which there were mainly leptons and their antiparticles.

THE LEPTON ERA

When the temperature was down to 100 billion degrees about one ten-thousandth of a second after the big bang the universe entered the lepton era. It was now a thick soup of radiation (photons) and leptons—mostly electrons, positrons, neutrinos, and antineutrinos. Again there was a thermal equilibrium with electron–positron pairs being created at about the same rate as they were being annihilated. But there were still a few protons and neutrons left from the hadron era—about one per billion photons. Neutrons in the free state, however, decay to protons and electrons in about 13 minutes. This means we had another important reaction going on, namely neutron decay. The temperature at the beginning of this era was, however, still high enough that an electron hitting a proton would create a neutron so we had a state of equilibrium. But when the temperature dropped to 30 billion degrees the electrons did not have enough energy to create neutrons, so the neutrons began to decay in large numbers.

Another important event that occurred during the lepton era was the decoupling and release of neutrinos. In the reactions involving protons and neutrons, neutrinos and antineutrinos are created. While the temperatures were sufficiently high they remained coupled to these particles, but when it lowered below a certain critical value they were released and expanded freely into the universe. And as the universe expanded they cooled until their temperature reached about 2°K. We have not yet detected them.

THE RADIATION ERA

A few seconds after the big bang, when the temperature was approximately 10 billion degrees the universe entered the radiation era. At the beginning of this era many leptons were still present, but when the temperature dropped below 3 billion degrees, the threshold for lepton pair production, they rapidly disappeared with the release of numerous photons. The universe then consisted almost entirely of photons.

Within the radiation era an event of extreme importance took place: the synthesis of the first nuclei. This is, in fact, that event that Gamow struggled with, discussed earlier. When the temperature was 1 billion degrees, about 3 minutes after time zero, the universe was cool enough so that when a neutron and a proton collided they stuck together creating a nucleus of deuterium (a heavy variety of hydrogen). The collision of two deuterium nuclei then created a helium nucleus. Thus, within a very short time, perhaps 200 minutes, about 25% of the matter in the universe was converted to helium. There is, in addition, a conversion of hydrogen to helium in stars, but it is only about 1% as great as this. Other elements were also produced during this era: small amounts of tritium and lithium, but the creation of nuclei beyond lithium was prevented. It might seem strange that we can talk about these events in such detail, and of course everything I have said so far is theoretical, but our theory seems to be in excellent agreement with observation, so we are encouraged. According to the theory, for example, about 25% of the matter in the universe should be helium, and observationally this is so.

THE COSMIC BACKGROUND RADIATION

The universe continued to expand and cool for thousands of years. At this stage it consisted mostly of radiation but there were some particles present: neutrons, protons, electrons, neutrinos, and nuclei of simple atoms. It was a boring universe, though, with little happening, and it was still opaque, as if filled with a thick luminous fog. The opaqueness was due to the equilibrium between the photons and the matter; the photons were, in effect, locked to the matter. Finally, though, at a temperature of 3000°K the electrons and protons combined to create hydrogen atoms and photons were able to break free from the matter. They decoupled, just as the neutrinos had earlier, and spread freely out into the universe.

It must have been like a miracle: the thick fog suddenly lifted leaving a transparent universe. But the temperature of the radiation was still high—just below 3000°K—so the universe was still red hot. It continued to cool, though—1000°K, then 100°K, until it reached its present temperature of only 3°K.

Gamow predicted in 1948 that this radiation should exist, but he made many errors along the way to his prediction, both numerically and in the details of the reactions. A few years later, however, his student corrected them and predicted a current temperature of about 5°K. They were sure, though, that this radiation would be impossible to detect—probably masked by starlight. Because of their uncertainty, 17 years passed before it was actually detected.

In the early 1960s a special radio telescope capable of detecting microwaves was installed at Holmdel, New Jersey, by Bell Telephone. It was used for communicating with the satellite Telstar. The two scientists assigned to it, Arno Penzias and Robert Wilson, planned on using it to study the microwave emission from our galaxy.

Before they could begin the study, though, they had to identify and eliminate any noise that might have been coming from the telescope itself or surrounding (terrestrial) sources. They decided to concentrate on a wavelength of 7.35 centimeters, but soon found there was a distinct hiss at this wavelength. They were certain that with a little work they could eliminate it, but to their surprise it would not go away—regardless of what they did. It became such a nuisance that they tried to find out if it was coming from the sky. Strangely it seemed as if it was; yet it seemed to be coming from everywhere. Regardless of where they pointed their instrument, the hiss refused to disappear.

Unknown to them a short distance away at Princeton University, two physicists, Robert Dicke and Jim Peebles, were discussing the possibility that there was a residual radiation in the universe—left over from the big bang explosion. Peebles calculated its temperature to be somewhere around 5°K and they encouraged two colleagues, P. G. Roll and Dave Wilkinson, to look for it. They were apparently unaware that Gamow had made the prediction many years earlier.

Penzias heard of the search and phoned Dicke suggesting that what he was looking for might be the "noise" they had found. Dicke visited Holmdel and within a short time it was established that the noise was likely the radiation they were searching for. They then proceeded to publish the results without any mention of Gamow and his student's earlier work. But when Gamow saw the paper he wrote an irate letter to Dicke telling him of the prediction. Penzias and Wilson were later awarded the Nobel prize for the discovery.

Of course, further proof that the noise was indeed the cosmic background radiation was needed. Penzias and Wilson had only got a single point at a wavelength of 7.35 centimeters on a radiation curve. Earlier we saw that any body with a temperature above its surroundings radiates, and its radiation curve (plot of the amount of radiation emitted at various wavelengths) has a characteristic shape. If the body is a perfect absorber, this curve is called a black body curve. As you go from long wavelengths to shorter ones, the curve rises, peaks, then drops off sharply: According to calculations, the temperature curve for the cosmic background radiation had to exhibit these features—in short, it had to be a black body curve.

The Penzias–Wilson point was the first on the curve but shortly thereafter Roll and Wilkinson obtained a second point. Then as the news spread, many scientists made measurements at different wavelengths, and numerous points were obtained. There was, unfortunately, a difficulty: all the points were on one side of the peak. It was therefore important to get points beyond the peak to be sure the curve turned over. In this region, however, our atmosphere hinders us: it does not let these wavelengths pass through and we therefore have to get above it to make the

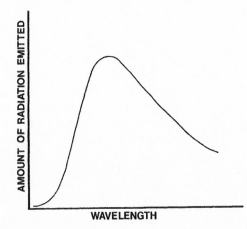

A radiation curve. If the cosmic background radiation was from the big bang, it had to fit a curve of this shape.

necessary measurements. But when a rocket was sent up to measure it, the scientific community was shocked: the point obtained was well above the curve. Scientists sighed in relief, though, when they discovered the detector had accidentally picked up some of the heat from the exhaust of the rocket. Later measurements showed that the curve did indeed drop off as predicted. We are therefore relatively certain now that this radiation is left over from the big bang.

To a first approximation, the radiation appeared to be the same in all directions, i.e., it was isotropic. But would it remain isotropic when more accurate measurements were made? Indeed, what would it mean if it were anisotropic (different in different directions)? If you stop for a moment and think about it, you soon realize that if the temperature of this radiation is slightly higher in one direction than in other directions, it means we are traveling through it in the direction of higher temperature. It is like being in a fog: if we measure the fog's density (thickness) around us and find it is thicker in one direction, we know we are traveling in that direction; correspondingly the density is less than the average directly opposite this direction. The first measurements in 1969 and 1971 hinted at some anisotropy so two groups, one from the University of California at Berkeley and one from Princeton, started preparing for more elaborate measurements above the atmosphere.

The University of California group, using the services of a U-2 spy plane, took the first measurements in 1976. And sure enough there was a small but distinct anisotropy that told us we were moving in the direction of the constellation Leo with a velocity of about 600 km/sec. It was later discovered that it is not only our solar system that is moving in this direction relative to the rest of the universe, but our entire galaxy and some of the neighboring galaxies are also moving in this direction.

THE GALAXY ERA

When decoupling occurred and the radiation broke free, the universe still consisted of a roughly uniform mixture of particles and radiation. The matter that would finally give us the galaxies was there but it still had a relatively smooth distribution. We know, however, that eventually it broke up (otherwise the galaxies would not have developed).

But where did the fluctuations that caused the breakup come from? Astronomers believe that they developed early, perhaps within the first fraction of a second after time zero. What caused them? We do not know for sure and, indeed, this may be a question we may never be able to answer with any confidence. Somehow they were generated almost immediately after the explosion. They may have been quite large at first, then smoothed out as the universe expanded, or they may have initially been small and grown. We still do not know which was the case. We do know, though, that as the radiation era ended, these fluctuations arose and began to grow. Over a period of time they broke up the expanding cloud of particles into individual fragments. These huge fragments continued to expand with the universe for a while, but eventually they broke free and began to lag behind. Then finally, as a result of self-gravity, they began to fall in on themselves. Most of them were spinning slightly at this stage, and as they contracted they spun faster.

The turbulence within each fragment was still so great at this point that further fragmentation occurred breaking up the cloud even further. This fragmentation continued until only star-sized clouds were left. These clouds then condensed into the first stage of a star, what we call protostars (the entire cloud at this stage is called a protogalaxy). Then one by one stars began to shine until finally we had galaxies similar to the ones we see around us today.

This may seem like a reasonable picture but many difficulties remain. For example: what did the earliest form of galaxies look like? These early forms are usually referred to as primeval galaxies, but we have not discovered one yet, so we have nothing to compare our theories to.

Still other problems existed. Consider for a moment what we see when we look into the depths of space. We know first of all that we are looking back in time; this is because of the finite velocity of light—it takes a certain amount of time for a beam of light from a distant object to reach us. A galaxy 10 million light-years away, for example, is seen as it was 10 million years ago; similarly, one 3 billion light-years away is seen as it was 3 billion years ago. As we look farther and farther out we see dimmer and dimmer galaxies until finally we see no more ordinary galaxies; past this point we see only what we call radio galaxies—galaxies that in many cases appear to be exploding. Also in this region and beyond we find particularly strange galaxies—strong radio sources that have exceedingly dense cores.

Finally, at the outermost reaches of the universe we see only quasars. Quasars were discovered in the early 1960s and have been an enigma ever since. They produce more energy than an entire galaxy (which contains several hundred billion stars) yet they appear to be small: no larger than the solar system. Their size seems ludicrous in view of their tremendous energy output. How could such a small object create so much energy? There has been considerable speculation over the years—much of it centering around black holes—but we still do not know the answer. The model that seems to be most acceptable is that they are a dense cluster of stars and gas centered on a gigantic black hole. Their energy is produced when the gas and debris are swallowed by the black hole.

An important point to bear in mind, however, is that when we are looking at these objects we are seeing them as they were long ago—when the universe was perhaps only a few million years old. Since we see only quasars here we are forced to ask ourselves if they were the first forms of galaxies. Furthermore, closer to us we have radio galaxies; perhaps quasars evolved into them. And finally we have ordinary galaxies; radio galaxies may have evolved into them. In other words, we may have an evolutionary sequence: quasars, radio galaxies, and ordinary galaxies. Although this view seems reasonable considering the evidence, most astronomers find difficulties with it and do not accept it. One of the major problems is the difference in size between quasars and galaxies. I should mention, though, that nebulosity has recently been found around some quasars. Perhaps this nebulosity condenses into stars eventually giving us galaxies. Because of this and other problems, most astronomers prefer to think that there are primeval galaxies out in this region but they are just too faint to be seen. Moreover, they have discovered some evidence to bolster this line of thought: recently, several galaxies have been found that are 2 billion light-years farther out than any known galaxies. They are so dim that it took 40 hours to get their image on a photographic plate.

REFLECTION

In the past few sections we have explored the details of the early universe: the freezing out of fundamental forces, the creation of the background radiation, the formation of galaxies, and so on. But how can scientists be sure their ideas or theories are correct? Obviously, they cannot just go to the telescope and look at the universe when it was a few seconds old. It is indeed a difficult task, but there is a way. There are things in the universe now that are a direct consequence of these early events. We refer to them as "relics." The major relics are:

1. The cosmic background radiation (with a temperature of 3°K).
2. The abundance of helium (approximately 25% of the total mass).
3. The homogeneity and isotropy of space.
4. The existence of galaxies implies fluctuations.
5. The matter–radiation ratio.

Ideally, the way things work is that scientists have a theory (based on the big bang in this case) that tells us a certain event occurs, say the release of radiation at 3000°K. Using our theory we follow this radiation through to the present; in the case of this radiation the theory predicts that it should now have a temperature of approximately 3°K. We then look for this radiation, and, as we saw earlier, we have found it. Similarly for helium: theory predicts that about 25% of the matter in the universe should be helium, and we observe a value very close to that. There are problems, though, with some of the relics; for example, we are still uncertain of the details of the fluctuations that caused the galaxies. And the standard big bang theory predicts there should be large numbers of magnetic monopoles (magnetic monopoles are particles possessing a single magnetic pole; magnets always have both a north and a south pole). So far, though, we have not observed any. Inflation theory solves some of these problems, but as we noted earlier it creates a few new ones of its own.

One of the main reasons for studying the very early universe is to learn more about unification, and its study has been extremely useful in helping us understand the problems related to this unification.

CHAPTER 7

Cosmological Controversy

The big bang theory has been around for about 40 years and is now generally accepted by most astronomers. You might think, on the basis of this, that it is problem-free, but this is not so. It does not answer a number of important questions and some of the answers it gives are not in agreement with observation.

Since our unified theory would obviously have to explain the structure and evolution of the universe, let us spend a few moments looking at its faults. First, it should be pointed out that the big bang theory is not a single theory; it is, in a sense, many theories, or at least it has many possibilities within it. It tells us the universe is expanding but it does not tell us exactly what caused the expansion or what will eventually happen to the universe. It does tell us indirectly, though, that it likely began as a gigantic explosion, and it tells us that it may continue expanding forever or it may eventually stop expanding and collapse back on itself, depending on how much matter is in it.

We do not know which of these will happen, but most of the evidence at the present time seems to point to the breaking point between them. In other words, the universe appears to be exactly on the dividing line between being open (forever expanding) and closed (eventually collapsing back on itself). This strange state of affairs is sometimes called the "flatness problem" since it means our universe, rather than being curved, is flat. It is something most astronomers feel we should be able to explain, but so far cannot.

Another problem has to do with the homogeneity of space. As we look out from earth we see that the universe (on a very large scale) is the same in all directions—the density of galaxies is generally uniform and the same types occur. Furthermore, the microwave background radiation is the same in all directions. If we think about this we begin to wonder how it is possible. If we assume the univeːse began, say, 18 billion years ago, there could have been no "communication" between galaxies that are now 20 billion light-years apart. The signal would have had to travel at greater than the velocity of light, and we know this is impossible. In short, the big bang was so powerful that some regions of the early universe were completely isolated from others, and as they continued to expand they remained isolated.

Our problem, then, is: if parts of the early universe were cut off from one another so that they could not communicate in any way, how did they all end up the same? This is something the big bang theory does not answer. And E. R. Harrison of the University of Massachusetts says, "We would feel more comfortable with this amazing state . . . if only we could explain why it exists."

The singularity is also not dealt with by the big bang theory. In fact, as we saw in the last chapter, we are not even certain a singularity ever existed. Furthermore, the big bang theory is written in the language of general relativity and we do not have a quantum version of general relativity. The main difficulty in trying to quantize general relativity (write general relativity in the quantum language) is that within the theory space is curved. In usual quantum theories such as the quantum theory of the electromagnetic field, we are dealing with flat space. Because of this lack of a quantum version of general relativity, we know little about the singularity or what happened immediately after the explosion. Furthermore, the big bang theory tells us nothing about what happened before the explosion. What, for example, was here before time zero? Was there another universe that collapsed, then suddenly exploded to give us our universe? This is obviously a good possibility, but we certainly have no proof that it happened. And there is a pretty good chance we never will.

Another difficulty centers on the origin of galaxies. According to Fred Hoyle, ". . . the hot big bang . . . is wrong because it cannot make galaxies in a decisive way." I mentioned earlier that they came about as a result of fluctuations in the gigantic gas cloud that originally permeated the universe. And there have been numerous papers written on how the event likely happened, but for the most part, they are speculative. Granted, some of the ideas look promising and that helps spur us on—nevertheless, we still have a lot to learn.

Also, according to the big bang theory, the universe should be full of magnetic monopoles (particles that have an isolated north or south magnetic pole). But so far, as mentioned earlier, we have not found a single one. There have been reports, but none have been confirmed. The inflationary modification of the big bang theory introduced in the last chapter overcomes some of these problems, but it has problems of its own. It overcomes the homogeneity problem, for example, in assuming that each isolated region of the early universe inflated so fast that each became an individual universe. This inflation also solves the monopole problem in that it dilutes the ones that were in the early universe by distributing them among many universes. Because of this there is now likely only a few in our universe.

But even with inflation everything is far from rosy. Scientists were attracted to Guth's theory in hopes that it would pull us through our problems. Unfortunately, when it was examined in detail it was found to be flawed. Still, it had so much going for it that scientists immediately began working on a revision, and early in 1982 a new inflation theory was published. It differed from the old theory mainly in that several contributions had been added to it from particle physics. Nonetheless, within a short time even this new theory was shot down: galaxies were not generated from the fluctuations that arose within it. So it's back to the drawing boards again. The theory does have promise, though, if all the problems can be overcome.

THE AGE OF THE UNIVERSE

Cosmology has also been plagued with another controversy. For many years the accepted age of the universe was taken to be approximately 20 billion years. This number appeared in most textbooks, articles, and popular books on cosmology, and most cosmologists accepted it since it was based on an extension of Hubble's work, done over a period of many years by Allan Sandage of Hale Observatory and Gustav Tammann of Basel, Switzerland.

But not everyone was convinced that Sandage and Tammann's number was valid. Gerard de Vaucouleurs of the University of Texas had also been working on the problem for many years using a similar technique, and he consistently got an age of about 10 billion years. Sidney Van den Berg, now of the Dominion Observatory in Victoria, Canada, also got a value close to this. For some reason, though, their results were ignored. Then in 1979 an announcement came from three astronomers that they had used an entirely different method and had got a value close to that of de Vaucouleurs.

Scientists finally began to take notice, and some began to wonder if the question should not be reevaluated. Most preferred to cling to the accepted value of 20 billion years, but as others got values close to 10, considerable controversy developed. Let us take a few moments to look into the details of this controversy.

We saw earlier that Hubble plotted the distance to galaxies against their redshift and was able to predict that the universe is expanding. Of particular importance in this plot is the slope of the line through the points; the value of this slope is called the Hubble constant (it is usually designated H). This constant is important because it is related to the age of the universe. It gives us a measure of the rate of expansion, and if we reverse this expansion, or what amounts to the same thing, if we reverse time (assume it runs backwards), the universe will contract. The age of the universe will then be the time it takes for all the matter to contract to a point. If the universe were expanding uniformly, its age would, in fact, be the reciprocal of H (one divided by H). But there is strong evidence that this is not the case; instead, the expansion seems to be decelerating, or slowing down. Thus, if we want to know the true age, we must take this into account, and to do this we must, of course, know how fast it is decelerating.

Using his "cosmic ladder," Hubble obtained a value for H in 1929 that corresponded to the embarrassingly low age of 2 billion years. It was embarrassing because geologists were getting an age that was much longer, and their evidence seemed pretty strong. But the embarrassment did not last long; Walter Baade of Mt. Wilson soon found a flaw in the way Hubble had estimated distances. He had used a relationship called the

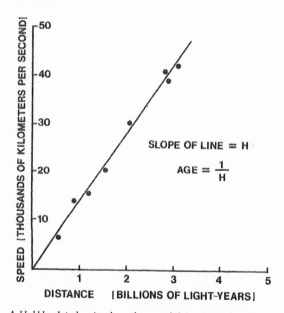

A Hubble plot showing how the age of the universe is calculated.

Cepheid period–luminosity law (the longer the period of a Cepheid, the greater its absolute brightness) to determine the distance to nearby galaxies, but the variables he used within these galaxies were not ordinary Cepheids and therefore did not obey this law. When adjustments were made, the age of the universe was doubled. Then a few years later Sandage noticed that Hubble had mistaken clusters, or groups of stars, for individual stars in his more distant galaxies. When the proper adjustments were made, the age was again doubled.

The accepted age as a result of this was approximately 10 billion years. But Sandage and Tammann were not satisfied. They went carefully through Hubble's work, redoing and extending it. At their disposal were valuable new techniques and methods of calibration, as well as the 200-inch Palomar reflector. The result of their work was another doubling of the age to 20 billion years. And for several years this number was gospel.

While Sandage and Tammann were redoing and extending Hubble's work, de Vaucouleurs was hard at work at the University of Texas. Like Sandage, he also had used a cosmic ladder, stepping his way out to dimmer and dimmer galaxies. But there was something that was bothering him. A few years earlier he had made a careful examination of the group of galaxies around us—called the Local cluster—and had established that it was part of a much larger group: a supercluster (or cluster of clusters). The dominant group within this supercluster was a giant cluster called the Virgo cluster (in the direction of the constellation Virgo). De Vaucouleurs became convinced that this supercluster was having an effect on our galaxy, and that was why he was getting a much smaller number than Sandage and Tammann. They were ignoring the effect.

But nobody was paying much attention to de Vaucouleur's explanation. Perhaps it was easier to believe we lived in an average region of the universe; de Vaucouleurs was saying that we were in an anomalous region. Obviously what was needed was an entirely different technique that could somehow resolve the controversy. The new technique (but not the resolution) came in 1979: Marc Aaronson of Steward Observatory, John Huchra of Harvard, and Jeremy Mould of Kitt Peak National Observatory announced they had obtained a value for H intermediate between that of de Vaucouleurs and Sandage. Most of their measurements, like most of Sandage's, however, were taken in the direction of the Virgo cluster. De Vaucouleurs suggested that they take some in a direction well away from the Virgo cluster. And—sure enough—when they did they got a value very close to that of de Vaucouleurs's.

Aaronson and his colleagues used a technique that had been developed a few years earlier by Brent Tully of the University of Hawaii and Richard Fisher of the National Observatory.

The Virgo cluster of galaxies. There are thousands of galaxies in this cluster. (Courtesy National Optical Astronomy Observatories.)

Tully and Fisher had determined the mass of a galaxy by observing its 21-cm line (spectral line emitted as a result of hydrogen in the galaxy). This line is broadened by spin, which meant that galaxies with the greatest spin had the broadest lines. And since it was well known that the largest, most massive galaxies rotated the fastest, Tully and Fisher merely had to measure the width of the line and they would have a measure of the "weight" of the galaxy, which in turn gave a measure of its true brightness, or luminosity. Once they had this luminosity, they could easily observe its apparent brightness and determine its distance.

The method seemed simple enough but it had problems in practice. First of all, galaxies are not usually seen face-on; they are usually seen at some angle and therefore much of their light is obscured by dust. A correction must be made for this. Tully and Fisher made the appropriate correction but their results were severely criticized.

Aaronson and his colleagues became interested in the method and decided to measure the infrared rather than the visible light that came from the galaxy. This way they would get around the corrections that Tully and Fisher were forced to make. Dust does not absorb infrared so they would not have to correct for the angular view of the galaxy. They eventually got a value for H that agreed with that of de Vaucouleurs's.

Aaronson and his colleagues soon became convinced that we are indeed living in an anomalous region of the universe. Because we are about 60 million light-years from the center of the Virgo supercluster, we are being pulled into this region with a high velocity. And if we were to get a correct value for the Hubble constant, we had to subtract this velocity from the measured recessional velocity of galaxies (velocity at which they are moving away from us).

Sandage and Tammann, on the other hand, are not convinced that we are living in an anomaly. Their measurements, they claim, show no such drift toward Virgo, and therefore they do not correct for it. Interestingly, the drift that Aaronson gets is not the same as that of de Vaucouleurs. Rather than falling directly toward the center of the Virgo supercluster, he believes we are spiraling in—his conclusion is based on a rather complicated rotating supercluster model.

Obviously we have a problem: Do we live in an anomaly as suggested by recent results, or are Sandage and Tammann correct? It would seem that we could easily answer this; after all, in the last chapter we mentioned that the universe is filled with background radiation, and we saw that its temperature is different in different directions. According to those results, we are moving toward Leo at approximately 600 km/sec; but Leo is about 43 degrees away from the direction of the center of the Virgo cluster! Thus, one set of measurements tells us we are moving in the direction of Leo, and another tells us we are moving toward Virgo. Which is correct? So far we do not know.

It appears, then, that we are stuck with the problem of the age of the universe: Is it 10 billion or 20 billion? There are, fortunately, two other ways to determine this number. Both actually give us the age of our galaxy, but we have a pretty good idea how much older the universe is than our galaxy so they are of considerable value. The first method uses giant clusters of stars called globular clusters; they surround our galaxy, like bees surrounding a hive. If we make a plot of the surface temperature of the stars in these clusters versus their absolute or true brightness, we get an interesting result. (This plot is called an HR diagram after its discoverers, Hertzsprung and Russell.)

Before I talk about this result, let us consider what the HR diagram of a cluster of stars usually looks like. If the cluster is relatively young, most of the points lie in a diagonal across the diagram, called the main sequence; there are, however, a few points in the upper right, and a very few in the lower left. The main sequence represents stars from small red dwarfs up to giant blue ones. One of the most important features of such a plot is that when a star gets old it moves out of the main sequence. The uppermost ones, corresponding to giant blue stars, move off first, and as the cluster ages more and more move off—

always from the top. This means that as the cluster ages its main sequence gets shorter and shorter. And of particular importance, the point above which there are no stars (called the "turning point") gives us an estimate of the age of the cluster.

If we look at the HR diagrams of globular clusters, we see that their turning points are almost at the bottom of the main sequence. This means they are exceedingly old, and in fact, they range from about 8 billion to 18 billion years old. On the basis of this, it seems that our universe must be older than 10 billion years.

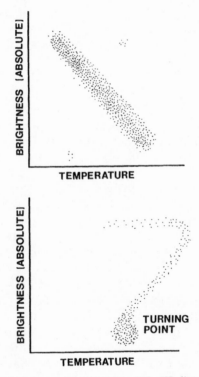

(Top) The HR diagram of a young cluster. (Bottom) The HR diagram of an old cluster, showing the turning point.

In the second method, scientists observe the decay rate of various radioactive materials. The unit by which this decay is measured is called the half-life of the substance—the time it takes one-half of the nuclei in a sample to decay. By measuring the half lives of radioactive atoms in the solar system we can determine its age, and from it we can get the age of our galaxy. The results again indicate that it is older than 10 billion years.

David Schramm of the University of Chicago and a number of colleagues have taken several of the methods of determining this age and incorporated them into a technique for "squeezing in" on the best age. They get a best age of about 15–16 billion years. But still everyone is not convinced. Harry Shipman of the University of Delaware has recently made a study of the number and evolution of white dwarfs in our galaxy. On the basis of his study he claims the Milky Way cannot be any older than 11 billion years. And Ken Janes of Boston University and Pierre De Marque of Yale agree with him. They have looked carefully at the way age determinations are made from luminosity–temperature plots of globular clusters and concluded that if errors in star observations and certain theoretical assumptions are taken into account, the age can easily be squeezed down to 12 billion.

And so it goes. All one can really say for certain at the present time is that the age of the universe is somewhere between 10 and 20 billion years.

THE REDSHIFT CONTROVERSY

It might seem as if things are in pretty bad shape if we cannot determine the age of the universe to better than a factor of two. To make things even worse, a number of astronomers have begun to question the observations on which the big bang theory is based, namely the Doppler shift of spectral lines that tell us the universe is expanding. According to the accepted interpretation, a shift toward the red means the object is receding from us, and the greater the shift, the greater the recessional velocity. On the basis of this we can, assuming we have an accurate Hubble plot, determine how far away an object is by measuring its redshift.

But what if we found two objects in the same cluster with different redshifts? Or if we found two objects that were obviously interacting and they had different redshifts. This, indeed, appears to be what has happened in recent years, and it has led a number of astronomers to question the Doppler interpretation. Leader among these mavericks is Halton Arp of Hale Observatories.

Upon obtaining his Ph.D. in astronomy, Arp went to work with the 200-inch reflector at Mt. Palomar. He soon became interested in peculiar galaxies. They are systems of one or more galaxies that are strange in some way: in some cases two or more galaxies are interacting, in others the galaxies have distorted or odd-looking filaments. He compiled a photographic atlas of these objects and published it.

His attention was directed one day to a peculiar galaxy that had several compact radio sources close to it. This caused him to look at some of the other peculiar galaxies, and he found that they also had compact radio sources around them (many of them were quasars). On the basis of this he put forward the hypothesis that the compact objects had been ejected from the peculiar galaxies. But most astronomers did not take him seriously.

As he continued his study he found even stranger objects. One of the strongest radio objects in the sky, NGC-520 (New General Catalogue number), had a line of four quasars emanating from it. Was it just a coincidence? Arp was convinced that it was not. Then he examined the system that eventually became the most controversial one. This system, which consisted of the galaxy NGC-4319 and the quasar Markarian 205, seemed to have a bridge between the quasar and the galaxy. The two objects appeared to be connected and yet according to their redshifts they were separated by millions of light-years: the galaxy was

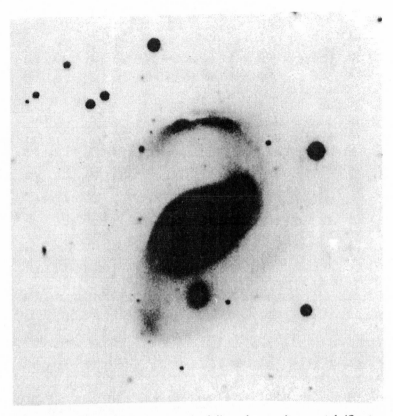

The Markarian 205–NGC-4319 system. Arp believes they may be connected. (Courtesy H. Arp, The Max Planck Institute.)

receding at 0.6% the speed of light, and the quasar at 7% the speed of light.

Was it possible? Most astronomers were convinced that the bridge was just a spurious effect resulting from the images of the two objects appearing side by side in the sky. They believed that in reality the quasar was far beyond the galaxy. But Arp photographed the combination in hydrogen α light and the bridge was still there. In fact, it was distinctly straight-sided, something that would not occur if it was just a spurious effect. If the quasar was far beyond the galaxy, there should be a distinct narrowing near the center—and there was not.

The first isophotal (a map of lines of equal brightness) photographs were obtained in 1974 by astronomers at Lick Observatory. They looked for a disturbance in the direction of the bridge and reported that they saw none. But they only looked at the outer faint isophotals, and Arp was not convinced. He, along with two colleagues, concentrated on the inner brighter isophotals and found that they were disturbed. The disturbance was not exactly in the direction of the quasar but Arp believes this is due to the galaxy's rotation since it ejected it.

And there are other objects. The galaxy NGC-7603 has a long luminous filament that terminates on a second galaxy, implying that they are somehow connected. Yet the smaller galaxy has twice the redshift of NGC-7603. Several clusters are also controversial. The Pegasus Quint, for example, was believed for years to be a closely knit group of five galaxies, yet when the redshifts were taken one of the group had a much lower redshift than the others, implying it was much closer. Yet a long luminous strand coming out of it appeared to be interacting with one of the other galaxies.

Arp has recently turned his attention to the southern skies and has published a Catalogue of Southern Peculiar Galaxies and Associations. Again, many anomalous cases showed up. One particularly interesting object is a three-armed spiral (most spirals are two-armed) with a large knot in one of the arms. Arp measured the redshift of the knot, more out of curiosity than anything he says, and to his surprise it had a redshift 4 times as great as the galaxy itself.

Is there anything to these discoveries or are they just coincidences? Arp tells us that we should keep in mind that there are now 38 discordant redshifted objects known to be associated

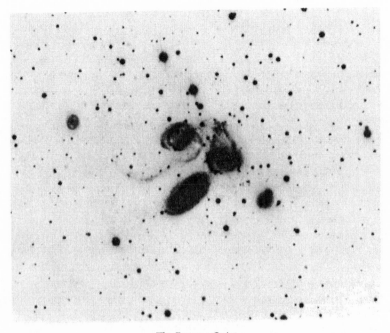

The Pegasus Quint.

with 24 different galaxies. This number, he claims, is large enough that we cannot just brush the problem off. He is convinced that, on the basis of this evidence, there is something wrong with the cosmological interpretation of the redshift (i.e., that the redshift gives recessional velocities that resulted from the big bang) in at least some cases.

Most astronomers, however, are not convinced by Arp's arguments. They believe that all of these objects will somehow eventually be explained perhaps as accidents or optical illusions, and our current interpretation of the redshift will not be affected.

RIVAL COSMOLOGIES

Because the big bang theory does not answer all questions of cosmological importance and perhaps because some astronomers like to leave the door open just in case, a number of rival theories have arisen over the years. Inflation theory is getting the most attention, though it deviates only slightly from the big bang theory in explaining the very earliest moments of the universe. Some of the rivals, on the other hand, differ significantly. I should point out, though, that just because there are a number of rival theories does not mean that most astronomers are dissatisfied with the big bang theory. This is not true; most are firm believers in it. Each of the rival theories has only a small group of adherents, and in most cases the rival theories have more problems than the big bang theory.

The first major rival to the big bang theory was put forward in 1948. The original idea came from Tommy Gold; he along with Hermann Bondi and Fred Hoyle had set up a discussion session on cosmological problems. At the time the big bang theory was not nearly as well established as it is now and Hoyle, in particular, was dissatisfied with it. If the universe was created in an explosion and was still expanding as a result of it, the galaxies should be thinning out, and there did not appear to be any evidence they were. Gold suggested that perhaps there was no explosion; he went on to explain his idea of a "steady state" universe, one in which matter was created between the galaxies. The universe had always been as it is today, and always would be in a steady state.

Thus, the steady state theory was born, and for a number of years it had a following. Hoyle, its strongest supporter over the years, is now one of the world's best known cosmologists, having written extensively on cosmology—at both the technical and the popular level. He grew up in England and went to Cambridge, staying on after graduation for many years. Besides his large contribution to the cosmological literature, Hoyle has written many delightful science fiction novels. One, *The Black Cloud*,

Fred Hoyle (to the left). (Courtesy AIP Niels Bohr Library, E. E. Salpeter Collection.)

has attracted considerable attention. It is the story of a cosmic cloud of gas and dust that has the ability to think and act. It has, interestingly, been made into a planetarium show. Hoyle is also an avid chess player and he has written a musical comedy.

Gold, the youngest of the three men, is now at Cornell University. Over the years he has worked in cosmology, plasma physics, radio astronomy, and planetary physics. In looking back at the discovery of the steady state theory, Gold recently said, "Most of my scientific education and insight came from my association with Bondi during the war. The three of us, Hermann Bondi, Fred Hoyle and I worked and roomed together. As there was little to do in the evening we discussed various cosmological problems . . . and Bondi, as the most able mathematician of the group, taught me dynamics and such."

Many of their discussions centered around cosmology but Gold admitted that at the time most of what he knew about cosmology came from the semipopular books of Jeans and Eddington. Hoyle would pace back and forth saying in a loud voice, "What does this Hubble observation really mean?" "Why is that necessarily true?" Bondi usually sat cross-legged on the floor, according to Gold, making rapid calculations as suggestions were hurled at him. At times he later admitted he was not even sure what he was calculating.

The association of the three men continued after the war, and it was in 1946 that they went to see a ghost-story movie (Hoyle thinks the name was "The Dead of the Night"). The plot was so arranged that the ending of the movie was the same as the beginning—leading, in essence, to a never-ending story. Gold was particularly impressed by it and later, according to Hoyle, asked them, "What if the universe was like that?"

But nothing came of the idea immediately. Over a year passed before they again talked about it. When Gold presented the idea to the group for the second time, Bondi chuckled and said, "I'll have it shot down by tomorrow." But when he looked into the mathematical details he was surprised. There were no inconsistencies, and the idea seemed to be mathematically sound. Both Bondi and Hoyle were convinced, though, that the idea would not work because new matter would be needed to keep the universe in a steady state, and that would violate the conservation of energy. But Gold pointed out that the big bang theory also violated the conservation of energy, only it did it in a fraction of a second rather than slowly over billions of years (as would be the case in the steady state theory).

To Bondi and Gold's surprise, Hoyle began working on the mathematical aspect of the theory on his own. This particularly surprised Bondi because he was not sure Hoyle was mathematically competent enough at that time to handle it. But Hoyle incorporated the idea successfully into general relativity and went on to introduce what he called "c numbers" (associated with the generation of new matter). By early 1948 he had a paper ready to publish and Bondi and Gold became alarmed. After all, the original idea had come from Gold, and besides, Bondi was not at all convinced by Hoyle's mathematical dabbling. Gold and Bondi were much more interested in the philosophical implications of the theory; Hoyle, on the other hand, was interested only in the mathematical implications.

Gold and Bondi began working in earnest when they heard that Hoyle would soon be ready to publish. But they needed something really tangible to make their ideas publishable. Bondi was tinkering with the numbers one day when he suddenly noticed that a slight adjustment of Hubble's constant would fit in perfectly with the steady state idea. He rushed and phoned Gold, who immediately exclaimed, "That's it!" It was exactly what they needed and they soon had the paper written.

In the meantime Hoyle had finished his paper and submitted it to the Physical Society of London; thus, it appeared at first as if he was going to beat them to publication. But to his surprise it came back rejected, with the suggestion that he submit it to the Royal Astronomical Society. Hoyle was thoroughly annoyed, for he knew the publication delay at the Royal Astronomical Society was running about a year and a half. Since he did not want to wait that long, he submitted it to the *Physical Review* in the United States. About a month later he received a reply that they would publish it if he would cut its length in half. Hoyle was annoyed again, believing that it would completely destroy the paper if it were cut so drastically. He therefore reluctantly submitted it to the Royal Astronomical Society.

Bondi and Gold's paper had by now already been submitted, and Bondi, a good friend of the president of the Royal Astronomical Society, managed to get it published quickly. He also used his influence with the president to get Hoyle's paper published shortly thereafter.

The generation of matter in an expanding universe was one of the major difficulties of the steady state theory. How did matter suddenly pop into the universe? The generation of matter meant that the universe—in effect, the galaxies—were reprocessing themselves continuously, and therefore had to have many different ages. Some of them would be extremely old. The big bang theory, on the other hand, had all galaxies generated at about the same time (about 15 billion years ago).

According to the generation rate, the average age of galaxies in the steady state theory should be about one-third the Hubble age, or 6 billion years. We know, or at least believe, our galaxy is about 12–15 billion years old, which does not rule out the steady state theory. But when we study the galaxies around us, we find that they are all about the same age as ours; certainly none is considerably older, and there do not seem to be any as young as 6 billion years.

This was perhaps the first piece of evidence against the steady state theory. But during the 1960s scientists discovered more and more evidence against it. A second major problem developed when astronomers noted that there seemed to be an overabundance of radio galaxies far out in the universe. Since we are looking back in time as we look outward, this seemed to imply evolution—consistent with the big bang theory and against steady state. At first the evidence was weak and Hoyle fought against it. But then quasars were discovered in the outermost reaches of the universe and the theory seemed doomed. The final, and most decisive, blow came with the discovery of the microwave background in 1965. This was a remnant of the big bang, and the theoretical prediction of its temperature agreed with observation.

Hoyle along with Jayant Narliker fought to save the theory by making modification after modification, but soon it had few supporters. The original simplicity of the theory was gone. But Hoyle did not give up. In the mid-1970s he introduced another modification based on Mach's principle. It was a strange and mind-boggling theory in which the universe was compartmentalized. Again Hoyle said there was no initial explosion; we are fooled into thinking this was the case because as we look out into space we see masses as they were many years ago. And according to Hoyle the masses of fundamental particles were far lower long ago than they are today. In fact, the farther we look the greater the decrease, until finally we come to the edge of our "compartment." This is what we interpret as the big bang explosion, but according to Hoyle it is just the place where mass goes to zero. He also believes that the fundamental particles shrink in size as they decrease in mass.

Of course he still had to explain the cosmic background radiation—and he did. He believes it is starlight that has leaked through from one of these other universes—in other words, from one of the other "compartments." General relativity, says Hoyle, is invalid only a long way away from a boundary, and since it is valid for us we must live quite close to a boundary.

VARYING *G* COSMOLOGIES AND COSMIC NUMBERS

One of the strong influences on Hoyle and his colleagues in the development of the steady state theory was a paper that had been published some years earlier by Paul Dirac. Dirac had shown that certain dimensionless ratios of fundamental constants of the universe were equal to 10^{-40}. He assumed there was some significance to this, but to keep things in status quo he was forced to introduce a varying gravitational constant G.

Some of these so-called "cosmic numbers" were actually introduced before Dirac worked on them. Eddington became fascinated with similar ratios late in life and introduced a unified theory called fundamental theory based on them (we talked about it briefly earlier). Most people are convinced now that he went off the deep end, however. Dirac's work, on the other hand, is still taken quite seriously.

Dirac graduated with a B.Sc. in electrical engineering from the University of Bristol in 1925. Upon graduation, however, he began to worry about his aptitude for electrical engineering; he wondered if he had perhaps made a mistake in selecting it. He finally decided he had and back to college he went, this time to St. John's College at Cambridge where he became a research student in mathematics. Before long, though, he began dabbling in physics. Then he learned of the important discoveries being made in quantum theory in Europe, and he was soon thoroughly enchanted with the new theory. His love affair with it resulted in an independent formulation that shed considerable light on the two apparently different theories that had been put forward earlier. Dirac showed that they were really just two different forms of the same theory. He remembered this breakthrough as one of the most exciting moments of his life.

In 1937 Dirac began working on his "cosmic numbers." His first paper on the subject, it is said, was written while he was on his honeymoon. These numbers are dimensionless ratios of fundamental constants of the universe: things like the charge of the electron (e), Planck's constant (h), the gravitational constant (G), and the speed of light (c). One set, those associated with the microworld, is called the N1 set. A second set, N2, is associated with the macroworld. The astounding thing about N1 and N2 is that they were both of the order of 10^{40}.

We immediately ask ourselves: Is there any significance to this, or is it just a coincidence? Somehow it seems as if it does have some significance, and if it does it may be a link between the microworld and the macroworld, and therefore between general relativity and quantum mechanics.

Another thing we must look at, though, is whether they are truly constants. Some of them involve L, which is related to the size of the universe, and they must vary, for we know the universe is expanding. Dirac believed the number 10^{40} was definitely a constant, however, and therefore if L varied, something else had to vary to keep it constant. He assumed this to be the gravitational constant G. On the basis of this, he formulated a cosmology, but it was soon found to be at odds with observations and had to be dropped.

The German physicist P. Jordan became intrigued with Dirac's work in 1947 and formulated a theory based on general relativity. He allowed for particle creation but again his theory was soon found to be flawed and had to be dropped. Brans and Dicke later developed a similar theory.

But what if G does vary? Would we be able to measure the variation? First of all, the variation would have to be extremely small or it would have already been noticed. In the case of the earth, if there was a decrease in G it would have been hotter and smaller in the past than it is now. Do we have any evidence of this? So far we have not found any. Some people have pointed to the large midocean ridges as possible evidence (the earth would have expanded and may have caused such ridges). Geologists, however, do not accept this, and there is now strong evidence that it is not a valid explanation.

Another thing that would be affected is the orbit of the moon. The period of the moon would have changed slowly. Again it would be difficult to detect, but with sufficiently accurate instruments a small change could be measured. T. C. Van Flandern of the Naval Labs believes he has detected such a change. He has considered records of occultations of stars by the moon over the last 30 years or so. It was about 30 years ago that atomic clocks first were used in assembling these records. Van Flandern has carefully subtracted all other known effects and believes he has a residue left that resulted from a small change in G.

ANTIMATTER AND OTHER COSMOLOGIES

We saw earlier that to every particle there is an antiparticle. To the electron, for example, there is the positron, or antielectron. And when a particle and its antiparticle are brought together, there is an annihilation with the release of one or more photons. In other words, there is a conversion of matter into energy.

We also saw that pair creation was particularly important in the early universe. At the extremely high energies that occurred there, particle pairs were created in profusion. Furthermore, when black holes evaporate they give off pairs—the smaller and hotter they get, the faster they evaporate. In short, it seems that the universe appears to like a symmetry between particles and antiparticles. This, of course, leads to the question: Does the universe presently have a symmetry between particles and antiparticles? The answer appears to be no. One place where we would expect to see a considerable amount of antimatter, if this were the case, is in cosmic rays—rays from space—and we find very little antimatter in these rays.

But what about entire stars or even galaxies of antimatter? Is it possible that they could exist? It is, of course, possible, as long as there is no matter around; if there were, it would soon be destroyed. If our earth, and our bodies, were made completely of antimatter, it would be the matter that would be troublesome, and likely rare. We cannot, in fact, tell if a galaxy is made out of antimatter just by looking at it. But since we have no evidence they are, and our theories tell us that matter should predominate in the universe, most astronomers assume that it does.

Several scientists, however, are not convinced of this. A number of cosmologies based on the idea that the universe consists of half matter and half antimatter have been put forward over the years. One of the best known of these is the Klein–Alfvèn theory put forward by Otto Klein and H. Alfvèn. They assume the universe was at one time a "metagalaxy," a huge sphere (perhaps a trillion light-years across) composed of equal numbers of particles and antiparticles that were well separated. Gradually this metagalaxy began to collapse as a result of self-gravity. There is a problem, though, even in this simple assumption: we do not know if matter and antimatter attract one another. It is possible that they repel. Anyway, the metagalaxy continued to get smaller and smaller until finally the matter and antimatter began to collide and annihilate one another. A tremendous amount of energy was released as a result of the anni-hilation but the collapsing material had gained so much momentum by this time it was difficult to stop. Finally, though, as a result of the energy buildup, the collapse stopped and a rebound occurred. The universe then began an expansion—the expansion we observe today.

The theory is interesting but it has many difficulties. It does not, for example, adequately explain the microwave background. And there is the question of the size of the metagalaxy and what it existed in.

There are several other rival theories besides the ones discussed above but most are not taken seriously at the present time and I will not discuss them, except to briefly mention the tired light theory. It assumes that the universe is not really expanding but that the redshift in spectral lines we see from distant galaxies is due to a "tiring" of the light as it passes through billions of light-years of space. There is very little support at the present time, though, for such an idea.

OTHER UNIVERSES

We saw in the inflation theory that there may be many universes besides ours—perhaps an infinite number. But whether or not they exist, from a practical point of view, means little to us. They are presumably forever cut off from us. We also saw that Hoyle introduced the concept in his compartmentalized universe. At the boundaries we were again cut off from these other universes.

The idea of other universes may seem strange, particularly in view of the fact we can never observe them, but it is not a new idea. It was discussed as long ago as 1961 by Robert Dicke. The scientific world was perhaps not ready for it at that time, however, as few people paid any attention to it. It was not until 1973 when a paper by Hawking and Collins reintroduced the idea that scientists began to take it more seriously. They suggested that there may be an infinite number of universes and that all started with different initial conditions.

In some cases the other universes would exist parallel with us in time; in other cases they would not. A popular version is a model in which universes branch off from existing ones giving rise to near replicas. Interestingly, the many-universe idea is not restricted to cosmology, but has a counterpart in quantum mechanics. A quantum mechanical model put forward in 1957 by Hugh Everett, then of Princeton University, has been getting a lot of attention recently. It was produced in an effort to get around certain problems related to causality. Again, universes split off a given one—each time there is a chance event at the atomic level. And each of these new universes splits in the same way, leading eventually to an infinite number of universes.

It is logical to ask if there is any connection between the quantum mechanical models and the cosmological models. They are indeed similar, which is in itself important because it indicates a link between the microworld and macroworld. But there are basic differences, and if the idea is to be viable these differences must be overcome.

I should make it clear that although the idea of other universes has attracted the interest of many scientists, it is by no means accepted dogma. It is only interesting speculation. But is it ever likely to be accepted? This is, of course, a question we cannot answer. Science takes some pretty strange twists at times, and ideas that seem completely outrageous to one generation are commonly accepted during the next. There are, indeed, a number of scientists who feel there is no way out of the many-universe concept—we will eventually be forced to accept it. But that is, of course, a question for future generations to decide.

We have seen that our usual theory of a single universe has a rival—the many-universe theory. We have also seen, in general, that cosmology is still far from an exact science; many aspects of our most accepted theory—the big bang theory—are still controversial and there are many important questions it does not answer. It does not, however, have a serious rival. The inflationary variation shows considerable promise for answering some of the questions and overcoming some of the difficulties, if it can be made to work.

CHAPTER 8

The Final Fate of the Universe

The ultimate fate of the universe is obviously an important part of our overall unified theory. This means it must incorporate Friedmann's theory, but it must go much further than it. Friedmann's theory merely tells us that the universe will either expand forever, or stop expanding and collapse back on itself, depending on its average density. It does not tell us exactly how this will be accomplished. We have ideas, of course, and we are convinced that some of them are correct, but in reality many of them are quite speculative.

Let us begin, then, by considering the two alternatives of Friedmann's theory. We can best understand them by considering an analogy. Suppose you throw a ball up in the air. You know that it will gradually slow down, stop, then begin to fall back to earth. The height to which it rises depends on how fast you throw it, and also on the strength of the gravitational field. If you had enough strength, you could, in fact, throw it fast enough so that it never returned to earth. The velocity that is needed to do this is the escape velocity we discussed earlier.

We have a similar situation in relation to the universe. Somewhere around 18 billion years ago the big bang occurred, creating the universe. Debris was thrown outwards at incredible velocities, and it is still moving outward in the form of galaxies. We do not have an earthlike object pulling the galaxies back to a common point as we did in the case of the ball, but we do have something similar: the mutual gravitational pull of all the galaxies. This pull slows the expansion of the universe, causing the galaxies to gradually decelerate. The outermost ones—which are farthest in time from us—should be slowing down the most.

The important question, then, is: Will the galaxies slow down enough so that they eventually stop? In other words, is the mutual gravitational attraction sufficient to stop the outward expansion? It is easy to see that this depends on the strength of the gravitational field, which in turn depends on the average density of matter in the universe (amount of matter per unit volume). Our question can therefore be reformulated as: Is the average density of the universe high enough to stop its outward expansion? We are, in fact, not able to give a definite answer to this question at the present time, but as we saw earlier it appears to be close.

Whether the universe will be open or closed depends on its density relative to a density of approximately 0.5×10^{-30} grams/centimeter (usually called the critical density). If the density of the universe is greater than this, it is closed and will eventually collapse; if it is less, it is open and will expand forever. It might seem, on the basis of this, that it would be easy to determine if it is open or closed: we would merely have to measure its average density and compare it to the critical density. Unfortunately, there are difficulties—serious ones. We can make a rather good estimate of the density of visible matter, but it is far from the critical density. Over 100 times as much mass would be needed to close the universe.

We know, though, that there is considerable material out there that we cannot see: small dim stars, dust, rocks, black holes, and radiation. Is there enough of this "invisible matter" to close the universe? At first glance it seemed as if there was not. Several studies during the 1970s—one by Gott, Gunn, Schramm, and Tinsley in 1974—concluded that the universe was open. But since 1980 important new discoveries have forced us to reconsider.

THE MISSING MASS

We refer to the additional mass that is needed to close the universe as the missing mass. The name is actually a misnomer because the mass is not necessarily missing; it may, in fact, not even exist. There are strong indications, though, that it may be there in a strange, unfamiliar form. We have known for years that there is a large amount of mass in galaxies that we cannot see—some of it associated with individual galaxies and some with clusters of galaxies.

Let us look at these two cases separately, beginning with individual galaxies. We can, with a little effort, determine the total mass of a galaxy. It might seem that the way to go about this would be to determine the average mass of the stars in it, then count the stars. But this is too difficult—if not impossible. To understand the method we use, consider the solar system. We know that the planets travel in orbits around the sun according to three laws discovered centuries ago by Johannes Kepler. One of these laws gives us the speed of the planet assuming we know the mass of everything inside its orbit (in the case of the solar system the sun has almost all the mass). We could, of course, turn this around and measure the speed of a planet and determine the total mass of the objects inside its orbit.

This also works for galaxies. Our sun, for example, is about three-fifths the way out from the center of our galaxy. By measuring its speed in orbit, we can calculate the mass of all the stars between us and the center. This, of course, does not give us the total mass of our galaxy; for that we would need a star at its outer edge.

Actually, we do not necessarily need a star—anything will do. Astronomers were, in fact, measuring the velocity of some of the outer hydrogen clouds in nearby spirals a few years ago when they noticed that they were moving much faster than they should have been according to estimates of the galaxy's mass. Upon studying the phenomenon further, they concluded there had to be a considerable amount of matter in the outer regions of these galaxies—a halo—that we could not see. Surprisingly, this halo was so extensive that it contained more mass than the stars.

What was the halo composed of? It could not be seen directly so it obviously was not ordinary stars. It might, however, be exceedingly dim stars, or perhaps rocks, dust, or gas. If all galaxies had these halos, they would obviously add a lot of mass to the universe. Was it enough to close it? When the calculations were made, astronomers found that we would still be far short. But this is not the end of the story.

Most galaxies in the universe reside in clusters; sometimes there are only two or three galaxies in the cluster but occasionally there are many. Our cluster, for example, contains about 30 galaxies. Once astronomers were able to determine the masses of individual galaxies, they turned their attention to these clusters. Adding up the masses of the individual galaxies in them, they found that there was not enough mass to keep them gravitationally bound. Yet they appeared to be bound—there was no indication the individual galaxies were flying off. Some of the clusters, in fact, needed several hundred times as much mass as they had to keep them bound. And even when the extra mass associated with the halo was added in, there still was not enough. It is perhaps easy to see, on the basis of this, why the universe has missing mass.

If there is indeed missing mass, we must ask ourselves what form it would take. It obviously has to be in a form that is not directly detectable or we would see it. One possibility is hydrogen gas, either neutral atomic hydrogen or ionized hydrogen (hydrogen that is charged after losing electrons). When we look at neutral hydrogen in detail, though, we find that it is not a good candidate. It gives off 21-cm radiation, and we see a 21-cm spectral line as a result. A study of this line from nearby and distant hydrogen has shown that there is little hydrogen in this form between the galaxies.

Ionized hydrogen was at one time thought to be a much better bet. Scientists knew that there was a background of X rays in the universe, which they believed was associated with the ionized hydrogen. It was later shown, however, that this background is likely due to quasars. Scientists then considered neutron stars, white dwarfs, and black holes. But they too were shown to be poor candidates. Black holes would have to be supermassive—of the order of a galaxy's mass—or else extremely common throughout the universe, and this seemed unlikely. Studies showed that although there may be massive black holes at the cores of many, and possibly all galaxies, there was little indication there were massive isolated black holes in clusters, and if they were common there should be some evidence of them in our galaxy.

Photons are also a candidate; they are, after all, energy and energy is just another form of matter. But again when the calculations were performed it was shown that they contributed very little.

It appeared as if there just was not enough matter in the universe to close it. But a few were convinced it would eventually be found. Then came what appeared to be the clincher. We saw in an earlier chapter that all the deuterium in the universe was made within a few minutes after the big bang. Though most of it was quickly changed into helium, a small amount survived. And measurements of this amount can tell us if the universe is open or closed. To see why, let us look at what was happening at that time. We know that the deuterium nuclei were colliding and helium was produced as a result. If the density of the universe was high, there would have been lots of collisions and considerable helium would have been produced. If the density was low, on the other hand, much of the deuterium would have survived. As the amount of deuterium in the universe has not changed much since that time, a measurement of it should tell us if the universe is open or closed. And, of course, measurements have been made. They indicate that the universe is open. This seemed to be pretty conclusive during the 1970s, and when similar measurements on helium gave the same result it seemed as if the universe was, without a doubt, open.

Within a few years, though, scientists discovered that there was a flaw in the argument. It implied only that the universe could not be closed with particles called baryons (heavy particles). Protons and neutrons are both baryons, so most of the things in the universe we are familiar with—stars, dust, hydrogen, and even stellar-collapse black holes—are made up of them. You might ask, in fact, if there is anything that is not. There are, of course, the nonbaryon particles: the leptons and the so-called "exotic particles." The leptons are too light to contribute significant mass but the exotic particles have been getting a lot of attention lately. The first of these particles to be considered was the neutrino, and for a while astronomers were convinced that it was the particle that was going to close the universe. Neutrinos are about as common as photons—about a billion to every atom of matter—and for years they were considered to have zero rest mass. They did, of course, contribute a certain amount of mass, since energy is just a form of mass, but as in the case of photons the amount was small and they would therefore be of little help in closing the universe.

Then in the late 1970s came the suggestion that neutrinos had mass. Though small, scientists theorized it would make a significant difference in their contribution to the overall mass of the universe. An experiment to check on the prediction was performed by F. Reines, H. Sobel, and Elaine Pasierb. They did not measure the mass directly as might be expected. It had earlier been discovered that there were actually three types of neutrinos: one associated with the electron, one with a heavier but similar particle called the muon, and a third with an even heavier particle discovered in 1977 called the tau particle. According to the theory, any of these types could change into another. In other words, they could oscillate back and forth in type, but they could only do this if they had a mass greater than zero. Reines, Sobel, and Pasierb set up an experiment to check on this oscillation and believed they found it.

When others attempted to verify the experiment, though, they could not. For a while it seemed that Reines and his colleagues had made a mistake. Interest in the experiment began to lapse when suddenly a team of Russians announced that they had measured the mass of the neutrino directly. But again problems developed. Several groups have tried to verify their results and so far none has. The issue remains unresolved.

Of course, even if the neutrino does not turn out to be massive there are the other exotic particles, and some of them are worth serious consideration. Associated with the gravitational field, for example, is a hypothetical particle called the graviton. Scientists have not yet detected gravitons but are convinced they exist. A theory called supergravity predicts that the graviton should have a partner called the gravitino; in fact, it predicts that all such particles should have partners: corresponding to the photon there should be a photino, and to the W particle, a wino. They are referred to, in general, as the "inos," and some scientists believe they are good candidates because they may be massive. But if they are not (or do not exist), we still have another candidate, which again presently exists only on paper. It is called the axion, and is quite different from the inos (much lighter). All of these particles are rather speculative at the present time, but they are being seriously considered.

Another particle that has drawn a lot of attention lately is the magnetic monopole. It is an extremely massive particle that has an isolated magnetic pole associated with it. If you are at all familiar with magnets, you will likely argue that this is impossible. You know, for example, that if you take a bar magnet and cut it in half, it still has a north and a south pole. If you cut it again, the same thing happens; in fact, it happens regardless of how many times you cut it. You just cannot get an isolated north or south pole. But Dirac predicted in the 1930s that such a particle existed. His theory spurred many experimentalists to search for it, but it was never found and gradually interest lapsed. Then in 1974 Gerard t'Hooft of the State University of Utrecht in the Netherlands, and independently A. Polyakov of the USSR, showed that monopoles are predicted by certain unified field theories. This caused a resurgence of interest and many searches were again initiated. One search was conducted by Blas Cabrera of Stanford University. After careful consideration he came to the conclusion he should be able to detect about one monopole a year, so he built the appropriate apparatus and waited. And his wait paid off. On February 14, 1982, his apparatus indicated he had detected the first monopole. The scientific world was excited, but there was considerable skepticism and because he has not yet detected a second one the skepticism has remained. Furthermore, other attempts to detect the monopole have failed.

One final candidate is also worth mentioning. We saw earlier that black holes were eliminated because they were created in the collapse of baryonic matter. But this applies only to black holes that were generated in the collapse of stars and we believe there are many black holes that were not generated in this way, namely the primordial black holes. Any that were here before deuterium was produced are still good candidates. This restricts us to relatively small ones but nevertheless it does give us another candidate. A further restriction comes from Hawking evaporation; Hawking has shown that all those formed with a mass less than 10^{15} grams have evaporated. This means the only ones worth considering are those with a mass from 10^{15} grams up to about 10^{32} grams. Since this is approximately the mass range of planets, they are usually referred to as planetary-mass black holes.

With all these possible contributions, it would seem that there would certainly be enough mass to close the universe. But David Schramm of the University of Chicago thinks not; according to the calculations he and his colleagues have made, however, it is extremely close. It is so close, in fact, that it almost seems exactly on the boundary between the open and closed universes.

OTHER METHODS OF DETERMINING IF THE UNIVERSE IS OPEN OR CLOSED

An exact determination of the average density of the universe is perhaps the best way of determining whether it is open

or closed, and in recent years it has been getting the most attention, but it is not the only method. Another is related to the
Hubble plot. If the acceleration of the galaxies is uniform, we get
a straight line out to the farthest reaches of the universe. But if
they are decelerating or slowing down, the line will be bent, and
we should be able to determine if it is bent enough to close the
universe.

Again the method seems simple. We merely have to make a
plot that extends to regions very close to the outer limits of the
universe and note the shape of the resulting curve. But as with
density measurements, there are difficulties. I have already
pointed out that it is very difficult to get accurate measurements
in deep space, and there are other problems. As we look farther
and farther into space, we are, of course, looking back in time,
and therefore seeing galaxies as they were long ago. This leads
to a problem related to evolution: How do these galaxies look
today? How have they changed? Many theories indicate that
galaxies (particularly ellipticals) were much brighter than they
now are. This would mean that we are actually underestimating
their distance. On the other hand, some theories predict that
certain galaxies may grow by gobbling up nearby galaxies. They
would therefore be much brighter now than they were in the
past, and we would overestimate their distances.

There is considerable evidence of evolution as we look to
the outer limits of the universe. Past a certain point we see no
more ordinary galaxies, only radio galaxies, and in the farthest
reaches we see only quasars. If we try to use these objects in
making up a Hubble plot, though, we find they are entirely
useless; they lie well off the line for ordinary galaxies. Furthermore, we do not know exaxtly what quasars are, so we can
hardly expect them to help us. Because they are so far out (and
so young), it appears as if they might be the first forms of galaxies, but few astronomers accept this.

Another method of resolving this question relies on what
are called number counts. Again, the idea is simple but the
results are controversial. You merely count the number of galaxies (or other types of objects) in a given direction out as far as

possible, then make a plot of the number at various distances. In this way you can determine the overall curvature of the universe, and of course if it is positively curved it is closed, and if it is negatively curved it is open. A uniform distribution of points would be expected in all directions and at all distances if the universe was flat. If it were positively curved, on the other hand, we would expect an excess of points nearby, and if it were negatively curved we would expect a deficiency nearby. An extensive study undertaken by Ohio State University in the 1970s indicated an excess nearby, and therefore a closed universe, but recent studies have not verified this.

Angular tests are also worth mentioning, but consistent results have again not been forthcoming. In this method the diameter of a given type of galaxy nearby is carefully measured. Then the same type of galaxy much farther out (at known distance) in the universe is measured. If the universe is curved, we will appear to get an error in measuring the diameter: we will overestimate it if the space is positively curved and underestimate it if the space is negatively curved.

FATE OF THE CLOSED UNIVERSE

Because the universe seems to be so close to the borderline, in discussing its final fate we are forced to consider both the open and the closed case. Let us begin by assuming the universe is closed. If this is the case, little will happen for perhaps 40 or 50 billion years. Galaxies will continue to expand away from one another as the universe grows larger. But eventually the outermost galaxies will stop as the expansion reaches its limit and the universe will begin contracting. Where there were once redshifts of spectral lines, there will now be blueshifts. At maximum expansion, most of the stars within the galaxies will have died and they will consist mainly of white dwarfs, neutron stars, small stars, and black holes, surrounded by swarms of particles—mostly photons and neutrons. Eventually, in perhaps 100 billion years, clusters of galaxies will begin to merge into one another; very few collisions of individual objects will occur, but in time the universe will be a uniform sea of clusters. Then the galaxies themselves will come together, and finally the universe will consist of a uniform distribution of stars and other stellar objects.

All during the collapse black holes will be forming and growing as a result of accretion and collision. The background radiation will also be heating up; eventually it will be almost as hot as the surface of the sun and stars will begin to evaporate. As they move through space they will leave vapor trails, and begin looking like comets against a fiery bright sky. But soon a diffuse fog will overcome everything and they will disappear; the universe will have become opaque, mirroring the same transition that occurred in the early universe. (We saw in Chapter 6 that the early universe was opaque until it reached a temperature of about 3000°K, then a sudden clearing occurred.)

As the universe continues to collapse in on itself, the rest of the story is, of course, the same as it was in the chapter explaining the early universe, but everything is now in reverse order. Temperatures continue to rise, and smaller and smaller intervals of time become significant. The galaxies finally evaporate and merge into a primordial soup of nuclei, then the nuclei disintegrate. The universe then moves rapidly through the lepton and hadron eras into chaos. In the hadron era the nuclei themselves disintegrate into quarks. The universe is tiny at this stage and consists only of radiation, quarks, and black holes. In the final fraction of a second, the collapse reaches a near singularity, then the "big crunch" takes place.

BOUNCE

We do not know what will finally happen when the "big crunch" takes place because we do not have a theory for the exceedingly high densities that occur just before the singularity, but we can speculate. Much of the speculation has centered around the idea of "bounce." This is the case when the collapse suddenly stops and another big bang occurs, i.e., the universe begins expanding again. One of the reasons that bounce was first introduced is that it appeared to get around what many astronomers consider a nasty problem: a beginning to the universe. If it bounced once, it likely bounced many times, perhaps an infinite number, and if so we would never have to worry about a beginning.

Unfortunately, when scientists examined the details they found that bounce does not solve the problem. Between each of the bounces the stars in the universe radiate a considerable amount of starlight, which is compressed into the near singularity. In fact, this starlight accumulates and this forces each oscillation to be longer than the previous one. If we look back in time at such a situation, we see that the oscillations get shorter and shorter until finally we are back at what we are trying to overcome—a beginning. According to calculations, at most about 100 cycles would take us back to the beginning.

There have been various attempts to get around this problem. Tommy Gold, for example, has developed a theory around the idea that time reverses at the point of maximum expansion. Radiation would then pour back into the stars and the universe would be rejuvenated. It would then merely move back and forth between collapse and maximum expansion with cycles of equal length.

John Wheeler has also introduced an interesting, but highly speculative theory. Using an idea developed by Hawking that the fundamental constants of the universe are lost when the density is high enough, he has shown that the cycle need not always lengthen. Because of the uncertainty principle, the values of these constants are lost as the universe is squeezed to near-infinite density. If it then bounces and reexpands, they may have completely different values. The lengths of the cycles under these circumstances will again change, but in a random way: some will be exceedingly long, others short.

FATE OF THE OPEN UNIVERSE

In the alternative to the closed universe, namely the open universe, the universe continues expanding forever. The major difference, therefore, as compared to the previous section is the times involved. We talked there about 50 and 100 billion years, but in this section times are so long that we will be forced to use indices. We will, for example, discuss times up to 10^{100} years. If you thought 100 billion years was beyond your imagination, you are really in trouble now.

The first events that took place in this case are, of course, similar to those in the case of the closed universe. Stars continue to age, and when they are old enough they bloat to red giants, then either explode or slowly collapse and die. Some of them will undergo collision with other stars before they die. Collisions are exceedingly rare, and only a few have occurred in our galaxy (at least in the outer regions where we reside) since its formation. But when we start talking about time intervals as long as a trillion and a trillion trillion years, many collisions will have occurred. Some of these collisions will merely knock the planets off into space, but some will be strong enough to knock stars into completely different orbits. And a few will knock stars completely out of the galaxy. In fact, if we wait long enough it will appear as if the outer regions of the galaxies have evaporated.

Stars that are not knocked out of the galaxy, but undergo collision will likely be absorbed by the core. The core will eventually become a huge black hole. Finally in about 10^{18} years most of the galaxies will consist of massive black holes surrounded by a halo of white dwarfs, neutron stars, black holes, planets, and various particles.

The next development is predicted by a current unified field theory referred to as the grand unified theory (GUT, for short). We will talk about it in detail later. This theory predicts that the proton will decay in about 10^{31} years. There are, in fact, several experiments now in progress trying to detect this decay in an attempt to verify the theory. According to GUT, protons should decay to electrons, positrons, neutrinos, and photons. This means that eventually, everything in the universe made up of protons or neutrons—and this includes everything except black holes—will have decayed to these particles. The universe will then be a mixture of these particles and black holes and it will stay like this for an exceedingly long time. Eventually, though, the smaller black holes will evaporate. But there is a problem with the larger ones. The background radiation at this stage is exceedingly cool, but it is still hotter than the large black holes. Eventually, as the universe continues to expand, however, it will cool until finally the surface temperature of the black holes is higher than it. They will then begin to slowly evaporate, and as they do they will grow smaller (this process takes about 10^{100} years). The universe will then be overcome with electrons and positrons, which will begin to move around one another forming huge "atoms." But gradually the positrons and electrons will spiral in and annihilate one another, leaving only photons. This will leave us with a universe that consists of nothing but radiation.

We have seen the fate of the universe for the case where it is open and where it is closed. We still do not know which of these will occur. Furthermore, if the universe does collapse back on itself, we do not know if it will bounce. A unified theory would have to answer these questions.

CHAPTER 9

The World of Particles and Fields

We have seen the difficulties that exist on the macroscopic level in the quest for a unified theory of the universe, and we have discovered that some of these difficulties are tied to the microscopic structure of the universe. Here we will probe further to see the difficulties that exist in trying to bring the microscopic world into a unified theory.

Let us begin with particles. If you are like most people, you probably think of elementary particles as tiny billiard balls—spinning billiard balls if you happen to know they have a spin. It turns out, though, that this is not a very accurate picture. Quantum theory tells us that elementary particles such as the electron have a wave associated with them, which means that if we could see them they would look like tiny smeared-out clouds. And strangely, if we tried to locate one exactly we would not be able to; we would only be able to give its approximate position.

At one time we thought the electron, the proton, and most other similar particles were elementary; in other words, we felt they were fundamental in that they were not made up of simpler or more basic particles. But eventually the number of these so-called "elementary" particles got so large that we began to wonder, "Are all of them truly elementary?" For that matter, what do we even mean by "truly elementary"? We could say a particle is elementary if it has no further substructure, but even with this simple and somewhat vague definition we run into trouble. Consider the electron again. Suppose we discovered that it was made up of more fundamental particles; we would then immediately ask what these more fundamental particles are made of. And if we managed to answer that, we would ask what the subparticles were made of—ad infinitum. Somehow there has to be an end to this process. One way to stop it is to define a truly elementary particle as one that has no dimensions, and therefore no substructure; it is, in effect, a point. Obviously, something with zero volume cannot have a substructure. We do, in fact, believe the electron is a point particle: no experiment has ever shown that it has a finite radius.

This is not the case for the proton, though. Experiments have shown that it has a radius of about 10^{-13} cm, and there appears to be a definite substructure associated with it. In 1968 a high-energy beam of electrons was projected at protons at the Stanford Linear Accelerator Center (SLAC). The experiment showed that the proton's charge distribution was not uniform, but seemed to have a distribution characteristic of tiny subparticles. We now call these subparticles quarks. The family of quarks and the electron family, called the leptons, are now considered to be the only truly elementary particles.

As we saw earlier there are two other leptons besides the electron, called the muon and the tau, and each of these has an associated neutrino. The proton is in the family known as the hadrons, which in turn is divided into the baryon and meson families. The particles of these families are not elementary; each is made up of quarks.

The interaction of these particles with one another takes place via fields. In an earlier chapter we talked about electric and magnetic fields and saw how Maxwell showed that together they form the electromagnetic field. But what exactly do we mean by a "field"? From a simple point of view, it is a quantity that is defined at every point throughout space and time. It is, in effect, a device for expressing how forces between particles (or objects in general) are conveyed.

The simplest type of field is the scalar field in which each point of space is given only a magnitude. A good example of a

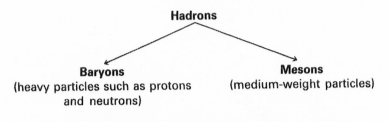

(Baryons and mesons are made up of quarks.)

Leptons

(light particles)

electron
muon
tau

(Associated with each type of lepton is a particle that is believed to be massless, called a neutrino.)

Fundamental particles of the universe.

field of this type is a region where the temperature at each point is specified (note, though, that this is not a force field). Another slightly more complex field is the vector field; in this case each point has both a magnitude and a direction. A good example of this case is a magnetic field; at each point there is an intensity and a direction (i.e., the direction of the lines of force).

For many years electromagnetic field interactions were visualized as due to action-at-a-distance forces. An electron, for example, passing close to another electron feels its electric field and is deflected. This is now referred to as the classical view. In quantum field theory a different approach is taken. We think of the force as being transmitted via an exchange particle. The exchange particle in the case of the electromagnetic field is the photon. When an electron passes close to another electron it exchanges photons, and it is this exchange that causes the de-

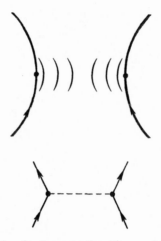

(Top) Two particles passing close to one another. The curved lines represent the electric field. (Bottom) A simple representation of the above interaction.

flection. In the same way, when we have a proton and a neutron or two neutrons very near one another we have an exchange of particles that causes an extremely strong force between them. The exchange particle in this case is different from that for two electrons.

In short, we have two fundamental entities in nature—particles and fields—and we are interested in how they interact. This is, in fact, what quantum field theory is all about. The first attempt to deal with this interaction from a quantum point of view was made shortly after quantum mechanics was formulated in 1926. But there were difficulties: the 1926 version of the theory could be used to quantize the particles but not the fields.

The first theory in which both the fields and the particles were quantized was introduced by Paul Dirac in 1927. He showed that the emission and absorption of photons by particles could take place quantum mechanically. But his theory also had a problem: it applied only to nonrelativistic particles, and we know that many of the particles of nature are traveling at velocities close to that of light when they interact. A relativistic theory was therefore needed.

DIRAC'S EQUATION

Again Dirac came to the rescue. He replaced the non-relativistic expression for energy with a relativistic one and derived an equation of motion. He soon found, though, that his equation applied only to particles with a particular spin. We know that the spin of elementary particles is not arbitrary; according to quantum theory it is quantized and therefore can take on only specific values. The spin of the electron, for example, can only have the two values ±½. (We frequently refer to the spin +½ as spins up and −½ as spins down.) Dirac's theory applied only to particles that had a spin of ½ and it was therefore a theory of the electron. It was, in fact, the first theory that had ever been derived in which the spin of the particle was predicted by the theory. But of more importance it was the basis of a complete theory of the interaction of light and relativistic particles.

Despite its successes, there were difficulties. It predicted that the electron could be in any of four possible states: two of spin (one with spins up and one with spins down) and two of energy (positive and negative energy). The problem was the negative energy states. If they actually existed, atoms could not remain stable. To see why, look at the diagram shown below that illustrates the various energies the electron can have—it is referred to as an energy level diagram. According to our theory, each level can contain only two electrons, one of each spin, and if there are any vacancies below that level one of the two electrons will drop down and occupy it. A photon will be released in the process.

But it is clear that the highest negative energy level is lower than the lowest positive energy level, and therefore a transition between these levels should be possible, again with the release

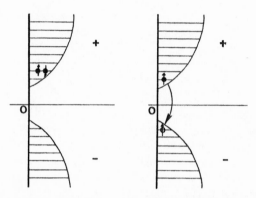

An energy level diagram. Energy levels at the top are positive, those at the bottom negative. The small dark circles represent electrons. The generation of electron–positron pairs is represented on the right.

of a photon. This means that any electron would find an infinite number of energy levels below it, and would cascade down through these levels, emitting photons at each step. The electrons in atoms would, in effect, be unstable and atoms could therefore not exist.

Several scientists looked carefully at this difficulty but it was Dirac who found a way around it. In 1929 he published a paper in which he postulated that the "sea" of negative energy states was filled. There was, therefore, no room in these energy levels for further electrons. His colleagues were skeptical; how could this be possible? We certainly did not observe the filled sea of negative energy electrons—and it should be all around us. Space should be filled with it. Dirac was undaunted by the skeptics, but the idea did apparently trouble him. Transitions from positive to negative energies were no longer possible, but there was still the possibility of a transition from the negative sea to positive energies. What would this look like?

Obviously this transition would take place if enough energy were supplied to the negative energy electron to elevate it to a positive energy, and calculations soon showed that the amount of energy was not excessive. The phenomenon should be observable. What would happen is that the electron, when raised to a positive energy, would leave a "hole" in the negative sea, and this hole would be observable. It would look like an electron in all respects except charge—instead of a negative charge it would have a positive charge. The only positively charged particle known at the time was the proton, and Dirac thought the particle had to be the proton. Oppenheimer pointed out, though, that it could not possibly be the proton; it had to have the same mass as the electron or else atoms would still be unstable.

What, in fact, would the process of creating a hole look like if we observed it? We would see an ordinary electron and a positive electron suddenly appearing at some point in space—a process we now refer to as pair production. Both of the particles would appear at once and both would be observable for a short period of time.

A few years later the process was actually observed by Carl Anderson at the California Institute of Technology while he was checking cosmic rays—particles and radiation generated in space—using a Wilson cloud chamber. In this chamber a cloud of mist is generated just after a particle passes through it; the particle, in its passage, generates ions (atoms that are missing some electrons) and droplets of mist form along the ion track, making it visible. If a magnetic field is used in conjunction with the chamber, a charged particle entering it will be deflected into a curved path (the direction depending on its charge). What Anderson saw was, in effect, a particle with the same mass as the electron that curved the wrong way. It traced out a path as if it were positively charged. Anderson called the particle the positron.

If the electron has an associated particle of opposite charge—an antiparticle—we are naturally led to the question: what about other particles? Does the proton, for example, also have an antiparticle? In turns out that all other particles have antiparticles. The antiproton was not discovered until 25 years later, however, because it takes considerably more energy to generate it than it does the positron.

One of the major things Dirac's equation did was change our view of the universe. At one time it was believed that the vacuum was filled with an ether—a mysterious substance needed to propagate light. But after Einstein formulated his special theory of relativity it was found that the ether was no longer needed, and the vacuum was considered to be empty. But with Dirac's theory it was again filled in the sense that if enough energy was supplied pairs of particles would be generated. Many different types could exist—space was filled with particles. But they had a brief existence and disappeared before we could see them. Space was obviously much more complicated than we thought.

Dirac was himself amazed at how much his equation predicted. He once said, "the equation is more intelligent than its author." His equation is now the basis of the interactions between electrons and protons via photons, a branch of field theory called quantum electrodynamics (abbreviated as QED). It is considered to be the most nearly perfect theory man has created, predicting phenomena to an extremely high accuracy.

THE INFINITIES

Despite the successes of Dirac's theory, many scientists still worried about the infinite sea of negative energy electrons. Dirac took the view, however, that it was the natural state of affairs and should not be worried about. I should emphasize at this point, however, that Dirac's view is just a way of interpreting the observations. We certainly do not see the absence of a negative energy electron in the laboratory; we see a positive electron.

Within a few years other infinities appeared that made the infinite sea of negative energies seem insignificant. To see how these infinities appear, I will begin by explaining how a field theory works (for now I will restrict myself to QED—the field theory of the electromagnetic interactions). It is based on what is called perturbation theory. In perturbation theory the interactions are treated in various orders: first order, second order, and so on. The largest contribution comes from the first-order calculations, second largest from the second-order ones, and so on—at least this is the way it is supposed to work. But when calculations were first made it was found that the first-order ones were in good agreement with experiment. There appeared to be no need for higher-order calculations so they were not made—besides, they were much longer and more complex. Eventually, though, Oppenheimer and Waller made the first higher-order calculation and discovered something strange. Instead of giving a small correction to the first-order terms as they should, they gave an infinite result. Waller mentioned this to Pauli, one of the foremost physicists of the day, but Pauli refused to believe it. Something had to be wrong; there was no way this could happen.

Let us take a moment to see why Pauli was so convinced. Consider, for example, the collision between two electrons; we can represent it simply as shown below. The point where there

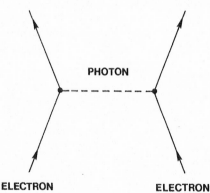

Feynman diagram of two electrons interacting. Photons are passed between them.

is an exhange of photons is called a vertex (marked by a dot). In the calculations there is what is called a coupling constant associated with each vertex. In the case of first-order calculations in QED, the magnitude of this coupling constant is $\frac{1}{137}$; second-order calculations also have the same coupling constant and should therefore be $\frac{1}{137}$ times smaller than the first-order calculations. But Oppenheimer and Waller showed that they were not—they were infinite. It was soon noticed that the difficulties seemed to be associated with the mass and charge of the particle, and also with the vacuum.

At first, scientists ignored the problem; after all, first-order calculations were in excellent agreement with experiment. Why would we even have to make higher-order calculations—besides, they were beyond our experimental technology. But then came the Lamb shift. The hydrogen atom had been studied extensively for years. The experimental spectral lines agreed well with the predictions of Schrodinger's equation, but when Dirac's theory was applied to it a hyperfine structure was predicted. It was extremely difficult to measure the fine splitting that was predicted, but T. C. Lamb and his associates succeeded in 1947 and the effect is now referred to as the Lamb effect.

It was a second-order effect and second-order perturbation theory was therefore needed to make a detailed calculation of it. This meant that the infinities that occurred in second order had to be dealt with. H. A. Kramers of Leiden University suggested that we should be able to make the calculation if the infinities were subtracted out in just the right way. But exactly how this was to be done was unknown. Lamb and N. Kroll made the first attempt, but they used an unreliable and crude method. Nevertheless, their method was moderately successful.

RENORMALIZATION

A good reliable method for subtracting out the infinities was obviously needed, and it came independently from three people at about the same time. The three were: Julian Schwinger, Richard Feynman, and Shin'ichirō Tomonaga. The first two were native New Yorkers, and the third Japanese. Schwinger was a child prodigy who entered college at 14, published his first paper in physics at 16, and had his Ph.D. completed by the time he was 21—quite an accomplishment even for a prodigy. He worked for a while under Oppenheimer at the University of California, but eventually went to Harvard, becoming a full professor while still in his twenties. He was a solitary person and preferred to work alone. While at MIT during World War II he usually worked in the labs at night when everybody else was gone. It was said that workers would occasionally leave uncompleted problems on the blackboard or on their desk overnight and find, much to their joy, that the solution had been filled in when they returned in the morning. His methods, unfortunately, are difficult to follow so we will follow Feynman's instead.

Feynman was quite different from Schwinger. He liked the high life and was frequently seen out on the city. He was seen so often at a certain topless bar that a reporter once asked him why he spent so much time there. He replied, "The atmosphere helps me think." His father had a strong influence on him, spending a large amount of time with him as a youth, instilling in him a love for nature and its workings. When he was about 13 he decided he wanted to learn calculus so he went to the library to check out a book on calculus, but the librarian refused to let him have it, saying that he was too young for such an advanced book. He wanted the book badly so he said that he was getting it for his father. The librarian apparently believed him and let him have it, and within a short time he had mastered it. He soon learned to his surprise, though, that his father did not know a thing about calculus. It was the first time he knew something his father did not.

He received his B.S. degree from MIT when he was 21 and his Ph.D. from Princeton University 2 years later. Shortly after he graduated he went to Los Alamos to work on the atomic bomb. "It scared me . . . the tremendous weapon . . . tremen-

Richard Feynman (1918–   ). (Courtesy AIP Meggerg Gallery of Nobel Laureates.)

dous potential," he later said. After it was successful he confessed to feeling guilty. "While I was celebrating the success of the bomb in Los Alamos by getting drunk . . . people were dying in Hiroshima."

Although the Nobel prize was awarded to him for his very "practical" work on renormalization, he has frequently said that he likes to do calculations just for the fun of it. And the problem does not have to be an important or practical one. He has, over the years, won many awards in addition to the Nobel prize, but he has a certain disdain for prizes. "I don't like honors," he said at a recent interview. "I already have a prize—the satisfaction of making the discovery and the pleasure of seeing others using it." Since 1950 he has been a professor of theoretical physics at the California Institute of Technology.

In developing his renormalization procedure, Feynman introduced a particularly useful diagrammatic technique for representing interactions. These diagrams could be used to write down the mathematical formulas for the process being considered. We have already used one of these diagrams (when discussing the scattering of two electrons). Another is shown below; it represents the emission and reabsorption of a photon (γ) by an electron (e).

But how, you might ask, is this process possible? It obviously violates conservation principles: initially there is just an electron present, then later both an electron and a photon exist—obviously the combined mass of the electron and photon is greater than the mass of the electron alone. This is true, of course, but it turns out there is a way around the difficulty. It involves the uncertainty principle—one of the cornerstones of quantum theory. This principle tells us that there is a certain fuzziness associated with nature on the microscopic level, and as a result of it there is an uncertainty in both the energy of the particle and the time when you measure this energy. You can, in

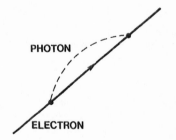

Emission and reabsorption of a photon by an electron.

effect, borrow a small amount of energy as long as you pay it back fast enough. It is like taking some money out of the bank and paying it back before your spouse knows it is gone. In a sense, it is as if it was never missing.

Because of this we can view an electron as continually emitting and absorbing photons, which means that it is, in effect, surrounded by a cloud of photons. We can, of course, never see, or directly measure these photons and therefore refer to them as virtual photons.

Now let us look at electron–electron scattering in more detail using this picture. Assume we have two electrons that pass sufficiently close to one another so that they are deflected. In QED we view this as an interaction between the two clouds of virtual particles. Some of the photons from one of the clouds move across the gap to the other cloud. We can consider an analogy to two skaters passing a softball back and forth to see why there is a deflection. The first skater throws the ball and in accordance with Newton's laws she is projected backwards, just as a gun recoils when you shoot it. The second skater is thrown back when he catches the ball, just as he would be if somebody gave him a push. The ball exerts a force on him.

There are, of course, numerous possible QED interactions, and each can be represented by a Feynman diagram. Another one, called the Compton effect (after the physicist who first studied it in detail), is shown on the next page. A photon (γ) is absorbed by an electron (e) at the lower vertex, then emitted a brief instant later at the upper vertex. In order to make a calculation using this diagram, we would obviously have to know the energy and momentum (a measure of the inertia) of both the

Simple representation of a cloud of virtual photons around an electron.

The interaction of two virtual photon clouds. It is similar to two people passing a ball back and forth.

The Compton effect. Absorption of a photon by an electron, and emission of it a brief instant later.

photon and the electron at the lower vertex. Our problem would then be to calculate the same quantities for the upper vertex. This is, in fact, the central problem of QED.

But let us look again at electron–electron scattering, which we represented as

Electron–electron scattering.

This is a first-order diagram, i.e., the diagram we would use if we wanted to calculate the process in the first order according to perturbation theory. As I mentioned earlier, though, there are second-order, third-order, and so on contributions to the process. Using Feynman's technique we can represent a typical second-order process as

A second-order Feynman diagram.

Others look as follows:

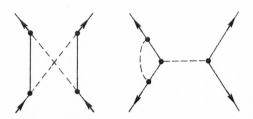

Other second-order diagrams.

There are, in fact, a large number of different second-order processes that can occur, and even more third-order ones. We can, in fact, now see why these diagrams should give a smaller contribution than the first-order ones. In two-particle interactions such as this, each pair of vertices contributes a factor $\frac{1}{137}$ to the calculation. And since we have two pairs the contribution should be 137 times smaller.

We can also now understand why we get an infinite result. Consider the charge of the electron; we can easily measure it and know it has a finite value. But in second-order perturbation calculations it appears to be infinite. If we think back to how we view the electron we soon see why. It is assumed to be surrounded by a cloud of particles, and this cloud screens its true charge. In the same way it screens its true or "bare" mass. The mass and charge that we observe are not, according to this view, the true mass and charge of the electron, but a combination of it and the effect of the screening. To get around this difficulty we must perform a subtraction. If the mass, for example, is made up of the observed mass and a "bare" mass (infinite), we must subtract off infinity. (A similar subtraction is made in the case of charge.) This procedure is called renormalization and the results that are arrived at using this technique are amazingly accurate.

But not everyone is pleased with the method. After all, infinity plus m(mass) minus infinity is not really equal to m. Why, then, does it work? Is it because we do not understand what we are doing? This is certainly true to some degree. We might even ask if perhaps we really have the wrong theory. And again we have to answer that we do not know for sure. We are, to some degree, sweeping things under the rug, but the important thing for now is that the theory works.

YUKAWA

QED was so successful that it soon became the model for the other interactions, namely the weak and strong, which were still not well understood. One of the first to use the analogy was a 28-year-old Japanese physicist who would have preferred to have become an experimentalist but found he could not master some of the techniques needed and so reluctantly went into theoretical physics. His name was Hideki Yukawa.

Yukawa assumed that if the electromagnetic force was due to an exchange of photons, the strong and weak forces must also be due to the exchange of a particle. But unlike the electromagnetic interaction with its infinite range, the strong interaction had a very short range; this meant that the exchange particle had to have mass. Yukawa used the uncertainty relation to calculate its mass, and he found that it was intermediate between that of the electron and the proton—about 200 times the mass of the electron. But there was no known particle of this mass.

At first, most scientists paid little attention to Yukawa's idea, but the following year (1936) a discovery was made that made them take notice. Carl Anderson, while making cloud chamber cosmic ray exposures on Pike's Peak in Colorado, observed a track that curved by the wrong amount. The curvature corresponded to a particle with a mass somewhat over 200 times the mass of the electron. Suddenly there was considerable interest in Yukawa's prediction. Was this the particle he was postulating? When they examined the new particle in detail, they found, much to their disappointment, that it did not interact with the nucleus. If it was Yukawa's particle (the exchange particle of protons and neutrons), it would have had to interact strongly. The particle was called the mu meson or muon, for short.

This left physicists in a quandry: if the muon was not the particle predicted by Yukawa, what was it? Why was it needed? And what about Yukawa's particle? Did it exist? It was almost 10 years before scientists found that it did indeed exist. In 1947 Powell of Bristol discovered another meson in cosmic rays, and this time there was a strong interaction with the nucleus. The particle was called the pi (π) meson or pion, for short. We now know that there are three pions: the charged pions $\pi^{\pm}$ and the neutral pion π^0.

It was soon clear that Yukawa's idea was credible and that the strong interaction was due to the exchange of mesons. We can represent it in a Feynman diagram as

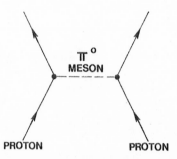

Feynman diagram of strong interactions. Two protons exchanging mesons.

Furthermore, just as the electron emits photons and absorbs them, so too does the proton (and neutron) emit pions and absorb them. This means that the proton and neutron must also be surrounded by a cloud of virtual pions. When two protons interact via strong interactions we can visualize an exchange of pions. There is, however, a major difference here as compared to the electromagnetic interactions: the range of the pion is about 10^{-13} centimeter, so the cloud is tightly packed around the particle, and if the two protons (or two neutrons, or a neutron and a proton) are to interact they must come within 10^{-13} centimeter of one another.

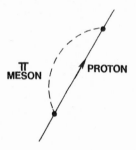

Emission and reabsorption of a meson by a proton.

The cloud of the proton is obviously complex in that it consists of both photons and pions. We can think of two protons exchanging photons when they are some distance apart (electromagnetic forces) but when they get extremely close to one another there is an exchange of mesons, giving rise to a much stronger force.

Yukawa also predicted that the weak interactions were the result of an exchange. The first important paper following up on this idea was by O. Klein in 1939. He called the exchange particle the W particle, a name we still use today. Julian Schwinger extended his ideas about 20 years later.

There were, however, difficulties with the above theories: they were not renormalizable. In addition, the coupling constant for the strong interactions was of the order of 1, compared to $\frac{1}{137}$ for QED. This meant that second- and third-order terms were just as large as first-order terms—even if we could renormalize. The coupling constant for weak interactions was much smaller, but there were other difficulties.

GAUGE THEORY

One of the important breakthroughs that took place about this time was the realization that symmetry played an important role in nature. We know that there is a symmetry between the electron and the positron: they are the same except for charge. Furthermore, there is an approximate symmetry between the neutron and the proton: they are the same except for the positive charge of the proton (and a slight difference in masses).

It was also discovered that symmetry was associated with invariance. We can best understand the relationship by looking at some simple examples (see figure below). Suppose we have a square; if we rotate it by 90 degrees it remains exactly the same; in other words it is invariant. Similarly a triangle undergoes an invariant transformation when we rotate it by 135 degrees. This also applies to the circle, but it is easy to see that it is the same at all angles. It is therefore said to have continuous symmetry. Furthermore, invariance is not restricted to geometry. Consider the field of charges shown below. For simplicity we will assume they have voltages of $+5$, $+3$, -10, -8 associated with them:

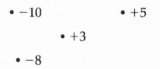

The voltage difference between the -10 and the $+5$ is 15 volts. Now imagine that we increase all of them by 50 volts: they will become 40, 55, 53, and 42. Again the difference between the two that were previously -10 and $+5$ is still 15 volts ($55 - 40$). By changing the background voltage we have left the voltage,

Symmetry of a square, a triangle, and a circle. Arrows show rotation. The object remains the same when rotated by the amount shown in the diagram.

and therefore the forces between the charges, unchanged. This type of symmetry is referred to as *global symmetry*.

The mathematics behind symmetry was developed by Evariste Galois, a mathematical genius who outstripped his teachers in high school, but because of his radical political views led a brief and tragic life. Bad luck seemed to plague him. Even as a teenager he was writing important papers in mathematics, but it seemed when he submitted them for publication they would always mysteriously disappear—much to his despair.

He got on the nerves of his teachers so much that he was

finally expelled. Then he got into a duel over a girl and was thrown in jail. As he sat in the damp dungeon, he became convinced he was going to die and he began writing out as many of his important discoveries as he could remember. It took him most of the night, but on the sheets he filled were the basis of group theory—a theory that has been invaluable to physicists in recent years. The next day Galois was shot and killed; he had not yet seen his 21st birthday.

The form of group theory that is used in modern physics is actually a slight modification of that due to Galois. The Norwegian mathematician Sophus Lie showed that the theory could be applied to continuous phenomena (phenomena that change very little over adjacent points), and he created an appropriate group theory.

The examples discussed above all have global symmetry, but it turns out that this is not the type that is particularly important in physics; a more important type is what is called *local symmetry*. To see the difference between global and local symmetry, consider the earth. We know if we displaced every city or town on the globe 100 miles to the right, things would not change; the distance between New York and Los Angeles would still be the same. If we displaced each city by a different amount, though, things would definitely change: the new distance between all cities would be different. In a local symmetry we make a displacement at each locality, as above, but the distances between the points (or differences of other types) remain the same. At first glance this may seem to be impossible. If we change each point differently it obviously has to affect the distance between points.

It was soon discovered, though, that the electromagnetic field (and therefore QED) had local symmetry—we refer to it as local gauge symmetry (the overall theory is called gauge theory). To see how this occurs, let us begin with a pure electric field. We know if we make changes at various points throughout it there will be an overall change in the field. In the case of the combined electromagnetic field, however, when we make a change in the

electric field there is an automatic change in what is called the
magnetic "potential" (associated with the magnetic field) that
compensates for it. Thus, the electromagnetic field has what is
called local gauge invariance (i.e., it remains invariant).

Scientists began to wonder if perhaps the weak and strong
interactions also had local gauge symmetry. When they were
examined in detail, though, it was found that neither appeared
to possess this property. At this stage, though, scientists did not
realize that local gauge symmetry was important; many believed
that it was just a strange quirk of nature. Of course, if neither
the weak nor strong fields possessed it, there was still another
possibility: maybe the combination of some of these fields, say
the weak with the electromagnetic, possessed it.

Indeed, maybe the electromagnetic and weak fields were
just different manifestations of the same field. Julian Schwinger
became convinced this was the case and assigned the problem of
discovering the link to his student, Sheldon Glashow. Glashow
arrived at a theory combining the two fields, but it was soon
shown to be flawed.

The breakthrough that eventually showed that Schwinger's
hunch was right came in 1954—although it was not recognized
as such at the time. Two physicists at Brookhaven, Chen Ning
Yang and Robert Mills, asked themselves what it would take to
make a global gauge theory into a local gauge theory. In considering the problem they dealt with a property of particles we call
isospin. The proton and neutron are thought to be two states of
the same particle that differ only in isospin: one state of isospin
corresponds to the proton and one to the neutron.

It was obvious to Yang and Mills that this system possessed
global symmetry, but they wanted to find out what they would
have to do to give it local symmetry. They found that they had
to add a new field. But once again there was a problem: the
exchange particle within the new theory had no mass (and the
exchange particle of the strong interactions did). This meant that
the technique could not be applied properly to either the strong

or weak interactions, and as a result there was little interest in the theory for many years.

THE WEAK INTERACTIONS

It was noticed during the 1920s that in certain reactions there was an energy imbalance: atoms emitting β particles (high-speed electrons) did so with less energy than they should have. Pauli suggested that an unseen particle was emitted in the reaction that carried off the missing energy. The following year Fermi named the particle the neutrino. It was believed to be a particularly elusive particle—chargeless and perhaps massless. And indeed it did prove to be difficult to find, but eventually (1956) it was found.

One of the most important reactions involving the neutrino is the decay of the free neutron; it decays to an electron, a proton, and an antineutrino in about 12 minutes. We refer to the process as β decay. The decay takes place via the weak interactions and therefore involves a W particle; it is, in fact, the best known weak field interaction.

The first theory of the weak interactions, or more explicitly, of β decay, was put forward by Enrico Fermi, an Italian physicist who emigrated to the United States shortly after Mussolini took over Italy. Although he is best known for his role in obtaining the first sustained fission reaction in a nuclear "pile," his contributions to physics were wide-ranging. His theory of β decay was quite successful, but it soon became obvious that much more was needed.

An important breakthrough in the field occurred in the mid-1950s with the discovery that parity (a symmetry operation in which we exchange right and left; mirror image of a process) is not conserved in the weak interactions. The conservation of parity had been taken for granted for many years when the Chinese physicists Chen Ning Yang and Tsung-Dao Lee began studying it. They were particularly interested in the decay of a particle called the K meson. At that time it was believed that there were two particles involved—one was called the τ and the other the θ. Yet, strangely, except for the decay itself, these particles had exactly the same properties. Yang and Lee considered the possibility that there was only one particle involved and looked at the consequences. This could be the case they found if parity was not conserved. At first this seemed preposterous—everyone knew that parity was conserved. But when Yang and Lee did a thorough search of the literature, they found that not a single experiment had ever been performed that proved this to be true. They published their conclusion in 1956. An experiment was carried out within a few months by Chien Shiung Wu of Columbia University verifying the prediction, and the following year Yang and Lee shared the Nobel Prize.

This was one of the first indications that symmetry could be broken, and scientists soon began to wonder if it was not broken in cases other than the weak interactions. One of the people who became interested in this possibility was Steven Weinberg, a physicist working at MIT. He learned about broken symmetry in 1961, and as he later said, "I fell in love with the idea . . . but was confused about its implications." The major thing that bothered him was that it involved a massless exchange particle, and the meson and the W particle both had mass. Despite this bothersome feature he spent almost 2 years developing the implications of symmetry breaking for the strong interactions. He tried to incorporate the Yang–Mills theory into his work but it too had massless exchange particles.

Then came the breakthrough: Higgs and Kibble in England showed that if the broken symmetry was a local gauge symmetry, some of the exchange particles (also called gauge particles) acquired a mass. This was possible if there was another, as yet unobserved, field associated with the vacuum. Weinberg was delighted when he heard about the idea and immediately went to work trying to incorporate it into his calculations, but at this stage he was still working on the strong interactions. Finally he

Steven Weinberg (1933–   ). (Courtesy AIP Niels Bohr Library.)

realized his mistake. He relates, "At some point in the fall of 1967, I think when driving to my office at MIT, it occurred to me I had been applying the right ideas to the wrong problem."

When he applied the new idea (now called the Higgs mechanism) to the combined electromagnetic and weak interactions, he found it gave three massive gauge particles and one massless one: this was almost exactly what he needed. Two of the massive ones would be the W, and the massless one would be the photon. But what about the one left over—it had mass and had to be neutral. Obviously there was another particle in the theory; it was called Z^0, and so far it had not been observed.

Let us take a moment now to see how the Higgs mechanism works. From a simple point of view, we can think of all the exchange particles as initially massless. The $W^{\pm}$ and the Z^0 particles then absorb Higgs particles and gain mass, and in the process the Higgs particles become ghosts. The photon does not absorb a Higgs particle and therefore remains massless. The process has been jokingly referred to as "the $W^{\pm}$ and Z^0 eat Higgs particles and get fat; the photon does not so it remains slim." Interestingly, the remaining unabsorbed Higgs particle is presently on the verge of detectability. Scientists assume that it will be found in the next few years.

With the completion of Weinberg's theory the weak and electromagnetic fields were unified into a single theory. Actually, at almost the same time that Weinberg was doing his work, a Pakistani physicist working in England, Abdus Salam, came to the same conclusion.

You might think that a unification of this type would send ripples of delight and excitement throughout the scientific community. But strangely it did not; in fact, the work was hardly noticed. Less than half a dozen people made reference to it in the following 4 years. The reason was that the theory was still not renormalizable. Both Weinberg and Salam were certain it could be made renormalizable, and both worked on the problem for several years, but neither succeeded in showing that it was.

The theory was beautiful but it could not be used to make calculations. Then one day in 1971 Gerard 'tHooft, a young graduate student, walked into Martin Veltman's office at the University of Utrecht and asked if he could work under him on a theoretical problem. "I want a hard problem . . . one nobody else can solve," 'tHooft said to Veltman. Veltman had indeed been working on an exceedingly difficult problem—the renormalization problem discussed above—and he had made some progress. Some of the diagrams that gave infinite results seemed to cancel, but not all of them; in fact, the overall problem was such a mess of complicated diagrams, all of which seemed to give infinities, that it would have been impossible to do them without the use of a computer. Veltman explained the problem to 'tHooft and asked him to work on it—not expecting him to make much progress.

A couple of months later 'tHooft walked back into Veltman's office and said, "I've solved the problem." Veltman looked skeptical, but listened anyway, and was surprised to find that 'tHooft's ideas were sound. He set the problem up on the computer according to 'tHooft's procedure and sure enough for each negative infinity there was a positive one: all the infinities canceled. The theory was renormalizable! Now scientists began to pay attention. Within a short time there was a rush around the world to dig up Weinberg's and Salam's papers of 1967. Calculations could now be made using the theory, and the predictions were in excellent agreement with experiment. Further verification of the theory came with the discovery of neutral currents—currents associated with the Z^0 particles predicted by the theory. In 1979 Weinberg and Salam, along with Sheldon Glashow, were awarded the Nobel prize for this work.

QUANTUM CHROMODYNAMICS

By the early 1950s the number of "elementary" particles had begun to get out of hand. Some, in fact, seemed much more like "resonances" than elementary particles. They acted like extremely short-lived excited states of existing particles. How did all these particles relate to one another? Was there indeed a relationship?

Among the newly discovered particles were K mesons and hyperons that appeared to be even more of an enigma than most of the others. They were created via strong interactions and like all heavy particles they decayed within a short time—less than a billionth of a second. This may seem like an incredibly short time, but to physicists it was too long: these particles should have been decaying in only a billionth of this time if they were decaying via the strong interactions. Their lengthy lifetimes meant that they had to be decaying via weak interactions—an exceedingly strange state of affairs. So strange, in fact, that they were called strange particles.

About 1953 Murray Gell-Mann considered the behavior of these particles. Like Schwinger and Feynman, Gell-Mann was a child prodigy, entering Yale on his 15th birthday and graduating from MIT with a Ph.D. when he was only 21. In 1955 he went to Cal Tech and was a full professor before he was 27. And unlike many physicists his education was not restricted to physics and mathematics; he speaks so many languages that he was introduced to the audience at a lecture I attended with the statement: "Dr. Gell-Mann is said to speak every language there is, including 'dog.'"

After a brief study of these new particles, Gell-Mann introduced the idea of "strangeness." Strangeness was a new number in quantum theory—a quantum number—similar to several that already existed. The strangeness number of neutrons, protons, and pions was zero; but particles like K mesons and hyperons had strangeness numbers of $+1$, -1, and -2. According to Gell-Mann this number is conserved in any nuclear reaction involving strong interactions. This meant that the total strangeness before the reaction had to be equal to the total strangeness after. Using this idea he explained the long lifetimes of the strange particles.

In 1961 Gell-Mann and Yuval Ne'eman, an Israeli physicist, used group theory to introduce a technique they called the "eightfold method" (it was based on eight quantum numbers). With this method they were able to arrange the elementary particles into families by plotting such things as strangeness and isotopic spin. These plots led to symmetric arrays of the particles such as those shown on the next page. The upper left one is the n–p family, which consists of the neutron, the proton, and the Σ, Λ, and Ξ particles. Some of the families had eight members and some ten, but all members of a given family had the same spin and differed only in charge and strangeness.

The patterns were, in a sense, similar to the periodic array of elements. Mendeleyev, in setting up the periodic table, did not know why the elements arranged themselves as they did (this was not discovered until some time later) but he did see that there were blanks in the table. This enabled him to predict the existence of several previously unknown elements. Gell-Mann and Ne'eman did not understand the significance of their

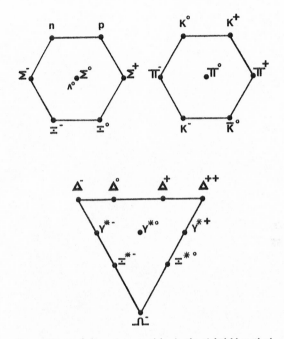

Arrangement of elementary particles in the eightfold method.

arrays either, and as in the case of the periodic table, one of their arrays had a blank in it (see peak in lower diagram). The properties of the unknown particle could be predicted and an extensive search was begun at Brookhaven in an attempt to find it. Over 50,000 photographs were taken, and in 1963 it was found. Many physicists had been skeptical of Gell-Mann's work up to this point, thinking it was nothing but an interesting juggling of numbers—not to be taken too seriously. But with the discovery of the new particle, now called the Ω^- (omega minus) particle, the scientific world took notice. It was a triumph of the new method.

But there was still a problem: why did the method work? There must be some sort of underlying scheme that was at the

Murray Gell-Mann (1929–   ). (UPI photograph, courtesy AIP Niels Bohr Library.)

basis of the method. Gell-Mann took up the problem in the early 1960s, but unknown to him another physicist, George Zweig, was also working on the same problem in Europe. Both men discovered that the eightfold method could be explained if hadrons were made up of more elementary particles. Zweig called these particles "aces," but the name suggested by Gell-Mann—"quarks"—is the one that is used today (he got it from a line in *Finnegan's Wake:* "Three quarks for muster mark").

According to Gell-Mann there were three types or "flavors" of quarks, which he called the up (u), down (d), and the strange (s). Each had a spin ½, and strangely, a charge not equal to that of the electron, but one- or two-thirds of it. And corresponding to each quark there was an antiquark, just as there are antiparticles corresponding to particles. All baryons were made up of three quarks; for example, the proton contained two up and one down quarks, and the neutron two downs and one up. Mesons, on the other hand, were made up of a quark and an antiquark. The π^+ meson, for example, consisted of an up and an anti-down quark.

$$(u)(u)(d) \quad (d)(d)(u) \quad (u)(\bar{d})$$
$$\text{p} \qquad\qquad \text{n} \qquad\qquad \pi+$$

The theory was beautiful; it completely explained the eightfold method and gave a recipe for every known particle. The third quark, called the strange quark, was important only in relation to strange particles. Ordinary particles were made up only of u and d quarks. Despite its success, however, there were problems with the theory. One of the major ones was that the quark had a spin ½ and therefore should obey what is called Fermi–Dirac statistics. According to this, only two particles can occupy the same energy state—one with its spin upward and one with its spin downward. But the formula for the Ω^- particle was sss, and all the s quarks had to occupy the same energy state.

The problem was overcome in 1964 when O. W. Greenberg

suggested that each of the three flavors of quarks came in three varieties. We now refer to these as "colors" (although there is no relationship to the usual meaning of color). Color is analogous to charge, or at least a quantity like charge (see table).

Quantum electrodynamics	Quantum chromodynamics
Electron	Quark
Charge	Color
Photon	Gluon

When the details of his suggestion were worked out, scientists resolved the statistics problem, and several other problems as well. In this new version, baryons were again made up of quarks of three different flavors, but this time each of the quarks had a different color. (Thus, quarks carry both color and charge.) It was important, though, that the three colors added up to white, for the overall particle had to be colorless. The mesons were again made up of a quark and an antiquark, one of which had to be the complement of the other so that they added up to white. Just as the charge of two opposite types gives a neutral particle, so too color plus anticolor gives a colorless particle. (Similarly, three particles can give a colorless result.)

It was soon discovered that color was more important than flavor. The fundamental triplet was color, not flavor (we will see later that flavor is no longer a triplet). With this discovery came a significant advance: the theory could be made into a local gauge theory. If it was a gauge theory, though, a new particle had to be introduced into the theory. It is now called the gluon (a pun—it is the glue that holds the nucleus together). Gluons are also colored, have a spin of 1, and, like photons, are massless but there are eight kinds of them versus one for the photon. Another difference is that gluons have color and therefore interact with themselves; photons do not have charge and do not interact with themselves.

Just as we previously visualized the electron as surrounded by a virtual cloud, we can now think of quarks as surrounded by clouds of virtual gluons. But things are a little different in this case. The quarks are assumed to be confined to a bag; a proton, for example, can be considered to be a little bag of two up and one down quarks. Each of the quarks in the bag has its surrounding cloud. In a proton–proton collision we can think of the two bags as approaching one another, then when they are sufficiently close there is an exchange of gluons from the two quark bags.

You may be quick to point out that there is a problem here. Previously we talked about the strong nuclear force as being due to meson exchange. So which is it: mesons or gluons? Scientists now assume that the true exchange particle is the gluon, but the exchange gives the appearance of the exchange of mesons. There is a close analogy between the way we view forces in molecules. We talk about covalent forces between atoms, but the covalent force is actually just the electromagnetic force.

One of the problems of the quark theory has been the confinement of the quarks and gluons. Why, for example, despite the many searches have we never seen either a free quark or a free gluon? To answer this we must go back to the bag model introduced earlier. Suppose we wanted to pull one of the quarks out of the bag. If we pulled on it a chain of gluons would be created according to our theory—a "string" would, in effect, be created with the meson on the end of the string. But strangely, the harder we pulled the greater the force that would pull back into the bag. This is opposite to the way the electromagnetic force works; when the two particles are close together, in that case the force is greatest. It weakens as they are separated. In the case of the gluon string the force is greater as the string lengthens. (This is what was observed in the SLAC experiments mentioned earlier; scientists found that the point charge appeared to be bound within the proton, and the greater their displacement the greater the force that bound them.)

But what if we did manage to break the string? According to the theory, a quark and an antiquark would appear at the break, thus the loose piece of string would consist of a quark and an antiquark (joined by gluons), which we know is a meson. In other words, if we tried to pull a quark out of a proton we would get a meson, and this is, of course, what is observed. If you bombard protons with sufficiently energetic particles, you get mesons out.

Although the color theory was highly successful, Sheldon Glashow did not like something about it. There were four leptons known at the time, and only three quarks had been postulated. He thought there should be a symmetry between the two kinds of particles.

Glashow's career parallels, in many ways, that of Weinberg's. The two were classmates in a Bronx high school; both received degrees from Cornell University and after several years both ended up at Harvard University. Despite the similarities in their careers, the two men have quite different personalities according to the people who know them. Whereas Glashow is an extrovert, Weinberg is quiet and reserved.

Glashow believed that there had to be a fourth quark associated with a quantity similar to strangeness. He called the new quantity charm, and the corresponding quark was called the charmed quark. The first evidence that this new quark actually existed came from two labs at almost the same time. It was discovered first by a group at Brookhaven working under Samuel Ting. But Ting was meticulous and spent a considerable amount of time verifying and rechecking his discovery. In the meantime a group on the other side of the continent at Stanford under Burton Richter also discovered the particle. What the two had discovered was not the charmed quark itself but a particle that consisted of a charmed quark and its antiparticle. Ting's group called the new particle J; Richter's group called it ψ (psi), thus, we had the ψ/J particle.

There was indeed a fourth quark, but before long another lepton was discovered, and if the symmetry was to remain it meant there had to be a fifth quark. Physicists called it bottom, and indeed within a short time it was found in a combination similar to that of the ψ/J, namely a combination of a bottom quark and its antiquark. This particle was referred to as upsilon. Since it no doubt has a neutrino, this also means that there is likely a further quark; it has been called top (it was discovered in 1984). In all, then, we have

Leptons		Quarks	
e	ν_e	u	d
μ	ν_μ	s	c
τ	ν_τ	b	t

How many more exist we do not know for certain, but it appears we may be near the end of the line. There are indications from cosmology that eight in each group is the upper limit.

SUMMARY AND FURTHER SUBDIVISION

We have seen that QED and the weak field have been brought together into a unification we call quantum electroweakdynamics (QEWD). Our next step is obviously to bring quantum chromodynamics into this scheme, and eventually also gravity, which at the present time seems to be completely outside of it. This full unification will be the subject of the next chapter. For now we would like to look at another question: Is the scheme of quarks and leptons outlined above the ultimate scheme? It might seem crazy to even ask such a question since I emphasized earlier that they are both considered to be dimensionless, and we cannot even isolate a quark to study it directly. Actually, what we have established in the case of the electron is that its structure, if it has any, is less than 10^{-16} centimeter. So far no experiment has shown any substructure larger than this. And it also appears we cannot isolate quarks, even at incredibly high energies. But that does not say they will never be isolated.

But if the theory of quarks and leptons is so good the way it is—it does in fact predict literally everything we observe—why look for an underlying simpler scheme? The first reason, perhaps, is the proliferation of particles. The scheme as originally set up had only 3 quarks; we now, including color, have 18, along with 6 leptons and there may be more. The number seems to be getting out of hand. But more than this, if you examine the theory in detail there are actually many unanswered questions. There are obvious symmetries within the theory that are unexplained. For example, everything seems to come in triplets: three leptons of charge -1, three of charge 0, three quarks of charge $+\frac{2}{3}$, three of charge $-\frac{1}{3}$. Why? And perhaps more important than this are what we call the generations of each particle:

$$\begin{pmatrix} u \\ d \end{pmatrix} \quad \begin{pmatrix} c \\ s \end{pmatrix} \quad \begin{pmatrix} t \\ b \end{pmatrix} \qquad \text{for the quarks}$$

and

$$\begin{pmatrix} \nu_e \\ e \end{pmatrix} \quad \begin{pmatrix} \nu_\mu \\ \mu \end{pmatrix} \quad \begin{pmatrix} \nu_\tau \\ \tau \end{pmatrix} \qquad \text{for the leptons}$$

These three generations are identical except for their mass. In other words, the μ acts exactly as if it were just a heavy electron, the τ an even heavier one. Also, the charmed quark seems to be just a heavy up quark. What is the relation between these generations? Are some just excited states of others? And even when you look further: why do the various particles have the masses they do? We have no explanation of the mass ratios between the various particles; they are large and unexplained.

Also there are the relationships between the charges. All colored particles have charges that are multiples of $\frac{1}{3}$ the charge of the electron; colorless particles are unit multiples of it. Why does nature favor this scheme over other possible ones? Again, this is unexplained within the theory. Also there is the relationship between electric charge and color charge that remains unexplained. And finally, is there a relationship between the quarks and the leptons? Do they really belong to one family?

A more fundamental theory need not invalidate our present theory; it might under certain conditions go over into it just as relativity goes over into Newtonian theory at low velocities, or quantum mechanics goes to classical mechanics outside the atomic region. Since our present theory is good down to about 10^{-16} centimeter, we would have to think of an underlying scheme that was within this scale—otherwise we would be able to observe it.

There has been a search for such a theory over the past few years. The object of such a theory would of course be to find a simple family of particles that contains fewer particles than our present scheme. In this scheme quarks and leptons would be made up of more fundamental particles. The second and third generations would be explained as excited states. Salam and Pati set up such a theory in 1974. Their fundamental particles were called preons, and with these preons they can build all the quarks and leptons. But the theory is less than satisfactory and not one family is needed, but three different ones. Hain Harari has also put forward a theory in which the fundamental particles are called rishons, but again there are difficulties.

If this approach were successful and we were able to find a single family and perhaps a single fundamental exchange particle, it would no doubt help us in our objective of finding a unified theory of the universe. In the next chapter we will turn to the present attempts to unify the microworld under the assumption that the quark and lepton families are actually one family.

CHAPTER 10

A Unified Theory of the Universe

We just described two theories of elementary interactions: quantum chromodynamics (QCD), the theory of strong interactions, and quantum electroweakdynamics (QEWD), the unified theory of the electromagnetic and weak interactions. Both of these theories are in excellent agreement with observation, both are gauge theories, and each is the crowning achievement of many years of work. The question that of course comes to mind now is: Can these two theories be unified into a single theory? It is important to emphasize that we want a unification of them more or less the way they are, and not a new theory that replaces them. Individually they are excellent theories, and we certainly do not want to throw them away. Since both are gauge theories, our new theory would also have to be a gauge theory.

Looking at the two theories we see that they encompass two families of particles: the quark family and the lepton family. And of course there is also a family of gauge particles that mediate the interactions between them. If we are to have a unified theory we must unify the quark and lepton families, i.e., show they are all basically the same (under certain conditions—not necessarily those that now exist) and we must also unify the gauge particles.

As you might expect this is quite an undertaking. We know, for example, that the leptons interact via the electromagnetic and weak fields whereas quarks interact via the strong and weak fields. In particular, leptons do not interact via the strong field but quarks do. If we are to unify them, both particles must interact in the same way.

What we would like to do is show that the two families are really part of a larger unified family. For this to be possible, though, we need a way to change a quark into a lepton and vice versa, but this can only be accomplished by introducing new particles. The first attempt to set up a theory that unified the two families in this way was by Howard Georgi and Sheldon Glashow of Harvard University in 1973. Since then many similar theories have attempted to do the same thing, but the Georgi–Glashow theory is still the simplest and the most satisfactory. It is a five-dimensional theory that has five basic particles in it, and like all such theories it is based on group theory. QCD, by comparison, is three-dimensional and has three colored quarks within it as basic particles. The Georgi–Glashow theory is referred to as grand unified theory or GUT for short; it is based on the group SU(5) (SU is an abbreviation for unitary symmetry; the numeral refers to its dimension).

The five basic particles within the new theory include three quarks of three different colors, the positron, and a neutrino. In addition to these particles of matter, there are 24 gauge particles that mediate the forces between them. You are already familiar with several of these particles: four are the particles of QEWD, namely the W^+, W^-, and Z^0 particles and the photon. Eight of them are the colored gluons of QCD, i.e., the particles that bind quarks into hadrons. This leaves 12 new particles; they are called X particles, and just as the gluons are the gauge particles of the color force, these new particles are the gauge particles of a new force, one we call the hyperweak force. They are colored and have spins of $+\frac{1}{3}$ or $+\frac{4}{3}$, and perhaps most important of all, they are the particles that convert quarks into leptons (and vice versa).

The introduction of X particles into the theory has an important implication: it means that a particle that we previously assumed to be stable—namely the proton—would now decay to two or more lighter particles. This may seem surprising at first but the proton is, after all, a relatively heavy particle and all heavy particles decay. Light particles such as the electron do not decay—but there is a reason. If a particle is to decay, it must decay to particles lighter (having less mass) than itself. The reason is that the conservation of mass must be obeyed: mass (or its energy equivalent) cannot be created or destroyed in any reaction. If the electron decayed to a particle heavier than itself, mass would have to be produced. Of course there are several particles lighter than the electron: the neutrino, the photon, and the graviton. Yet we have never seen an electron decay to any of these particles. Why? The reason is that the electron has a charge and these lighter particles do not; if the electron were to decay to one or more of them, one unit of electric charge would have to be destroyed and this is forbidden by the principle of conservation of charge. According to this principle, the total charge of the participants in any reaction must be equal to the total charge after the reaction.

Since a conservation principle stops the electron from decaying, we might ask if there is a conservation principle that does not allow the proton to decay. And indeed there is. It was formulated in 1949 by Eugene Wigner. He introduced what we now call the baryon number B; leptons and light particles have baryon number 0, and baryons have baryon number 1. According to his principle, the total baryon number of the participants in any reaction must be equal to that of the particles after the reaction. And until recently it seemed as if this principle was always satisfied.

As scientists looked more closely at the various conservation principles, though, they began to realize that some of them were more fundamental than others. Conservation of charge, for example, is a particularly fundamental one; it can never be broken under any circumstances. But others such as the conservation of strangeness are broken; it is not satisfied by the weak interactions. Was it possible that conservation of baryon number was also violated in certain cases? If so, protons could decay. Scientists eventually convinced themselves that this was the case. Indeed, it is not only proton decay that argues for its violation; another excellent argument comes from cosmology. It is well known that our universe is composed almost entirely of matter; at least there appears to be very little antimatter in it. Why? It seems more natural that the universe would be composed of half matter and half antimatter. We know that this is not the case at the present time but when the universe was first formed (the first fraction of a second after the big bang) it was. If we assume that baryon conservation is violated, we can show that the universe was initially symmetric, and the asymmetry we now see developed later. In other words, the theory explains, in a beautiful and consistent way, the present excess of matter over antimatter.

But if the proton did decay, what would it decay to? There are several possibilities: one is shown below, in which a quark is converted to an electron and a u quark to a $\bar{u}$ quark:

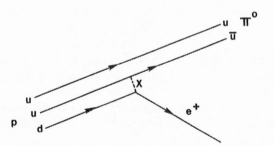

The decay of the proton to a meson (π^0) and a positive electron (e^+).

If we were to observe this reaction, we would likely see the creation of a positive electron (e^+) and a pion (π^0); the pion, however, would decay within a short time to photons (γ). The process would look like:

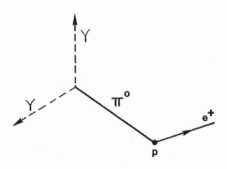

Creation of a pion (π^0) and a positive electron (e^+) with the pion decaying to two photons (γ).

There are also other ways the proton can decay. One of the u quarks could be converted to a d with the emission of an X; this X could then convert the d quark to an antineutrino. We would write this as $p \rightarrow \pi^+ + \bar{\nu}$.

Of course, if protons do decay the process must be incredibly slow. The reason is obvious: our bodies are made up of protons (also electrons and other particles) and if the rate was very high we would be radioactive. Even a small amount of such radiation would be disastrous to us; we would likely develop cancer in a short time. We know we are not radioactive—in fact, not even the slightest amount of radiation can be detected coming from our bodies. This tells us the lifetime or, more exactly, the half-life (time for half of all particles in a given sample to decay) of protons must be greater than 10^{16} years.

The first experiment that was set up in an attempt to determine the half-life gave a number much greater than this. The experiment was performed in a deep mine in India. Scientists determined that the lifetime had to be greater than 10^{30} years. Interestingly, shortly after this experiment concluded, Georgi, Quinn, and Weinberg showed on theoretical grounds [based on SU(5)] that the lifetime should be about 10^{32} years; they later reduced this to 10^{31} years. This is obviously an incredibly long time; our universe is only about 10^{10}, or 10 billion years old. Could we detect the decay if it were this long? The answer is yes—assuming it is no longer than 10^{32} years (beyond this point difficulties arise). We could do it by gathering a large number of protons together. If, for example, we had 10^{32} protons in a group, we would expect one of them to decay each year. Such a collection would not, in fact, be that large; it would of course depend on the particular type of material we were dealing with, but it would likely be no larger than an average room in a house.

I mentioned that the experiment in India was housed in a deep mine, and there was a good reason for this: the earth is being continually bombarded by cosmic rays; hence, we would not be able to distinguish many of the events they would create from proton decay. Also, it might seem that since a large number of protons are needed the experiment would be particularly expensive, but this is not the case. Protons are a basic constituent of all matter so we can use relatively cheap substances such as water, iron, or concrete. Experiments have, in fact, been set up using each of these materials.

In short, then, what is needed is a large mass of any material and a protected area to house it. In Europe there are many long tunnels that are ideal protective areas; those under Mt. Blanc have proven to be particularly useful as they have storage rooms off the main tunnel. Experiments have been set up in these areas. In the United States scientists have concentrated on mines. An experiment has been set up under Lake Erie in a Morton salt mine, another in a deep silver mine near Salt Lake City in Utah, and a third in an old iron mine in Minnesota. Concrete was used in the Minnesota experiment, and water in the salt mine experiment.

A heavy substance such as iron or concrete has an advantage over water in that the same number of protons takes up much less room. But there is also a disadvantage: the detectors must be placed relatively close to one another and this is difficult to do. In water, on the other hand, the detectors can be widely separated; in fact, they are usually arranged in a pattern around the edges of the tank.

Scientists are attempting to use an effect called the Cherenkov effect (named after the discoverer) in the water experiments. When a particle travels in water at a velocity less than that of light in vacuum but greater than that of light in water, it creates a cone of blue light. The cone points in the direction that the particle is traveling and the angle it subtends depends on its velocity. It is expected that the particles given off in proton decay will create Cherenkov cones, which can easily be detected.

Preliminary results from these experiments indicate the lifetime is slightly longer than 10^{32} years. This is in mild disagreement with the current theoretical prediction of approximately 10^{31} years, but the experiment is far from over. Scientists hope, however, that the lifetime is not too much longer than 10^{32} years (assuming it does decay) because neutrino events start to become common in this range, and may not allow us to see proton decay.

Although most physicists have their hopes pinned on the proton decay experiments as a verification of GUTs, it is not the only check on the validity of the theories. Some of these theories also predict a new particle called a magnetic monopole. We talked briefly about this particle earlier. Electric fields are generated by charges (and also by changing magnetic fields); an electron, for example, is surrounded by an electric field. Magnetic fields, on the other hand, are generated not by charges but by magnets, which have both a north and a south pole. There appears to be no such thing as an isolated magnetic charge of a single polarity (N or S).

Many scientists look upon this as a flaw in electromagnetic theory. It seems much more reasonable, considering the relationship between electric and magnetic fields, that there should be an exact symmetry in the way they are created. We should have a magnetic analogue to electric charge. In other words, we should have a magnetic monopole giving rise to a north magnetic field, and one giving rise to a south magnetic field. But nature does not appear to work that way. Why? Is it possible that magnetic monopoles exist but we have not found them yet? Many scientists believe this is the case.

Interest in magnetic monopoles began in 1931 when Dirac developed a theory that predicted them. But the particles that he predicted were not exactly what scientists were hoping for: they were strange particles in that they had a string attached to them.

The difficulty was overcome in 1974 when 'tHooft showed that monopoles are also predicted by GUTs, and his particles were quite different from those predicted by Dirac. They did not have a tail, but they were excessively massive—10^{16} times as massive as the proton. This was obviously why we had not seen them; there is no way we can create such a particle with present-day accelerators. Still, they should have been produced in the giant accelerator that nature provided—the early universe. According to GUTS they should have been created about 10^{-35} second after the big bang; particles of both polarity should have been generated and there should still be a residue of them left in the universe.

The prediction sparked an extensive search. Many groups around the world began searching for the particles and within a year one of the groups claimed they had detected them. (The claim has never been verified, though, and is now generally believed to have been a false alarm.) Cosmic rays were searched, moonrocks were searched, and there was even an experiment set up in Skylab to try to detect them. But none of the searches was successful.

If monopoles do indeed exist, where would we expect to find them? Because of their magnetic field, they would interact with the earth's field and be directed toward the poles. N monopoles would be attracted to the S pole, and S poles to the N pole. Also, it has been shown that they would probably travel much slower than originally believed, possibly much less than the velocity of light. In the hopes that some of them were still embedded near the poles, scientists had large blocks of ice cut from the regions and searched. But again there was nothing.

If they are so difficult to find, we might ask ourselves how abundant they ought to be. First estimates were that they should be about as common as protons. But of course if they were, we would have no trouble finding them. Later estimates reduced this number to about one per 10^{15} protons. And the number may actually be considerably less than this. E. N. Parker of the University of Chicago has pointed out that if they were very abundant, their field would interact with that of our galaxy and destroy it. Since this apparently has not happened, there may be far fewer than we suppose. Still, the search goes on.

Earlier we referred to another particle that has a tremendous mass and also existed in the early universe—the X particle. As in the case of the monopole, we have not detected X particles and the reason again is their huge mass. With the X particle, though, there is a further problem: it is a gauge particle and if it is to be unified with other gauge particles it must, under certain conditions, have zero mass. Why, then, is it so massive now, assuming it exists? To answer this it is best to look back at how the W particle gained mass. Above a certain temperature in the early universe, the W particle and the photon were massless. This temperature corresponded to an energy of 100 GeV (giga-electron volts = 1000 million electron volts). As the universe expanded and cooled through this temperature, however, the W's suddenly gained mass through spontaneous symmetry breaking. We say they ate Higgs particles and gained weight. Below 100 GeV the W's have mass and the weak and electromagnetic are distinguishable forces of nature. They are said to have "frozen" out of the unified field. As a simple analogy we could consider a homogeneous mixture of three liquids. Above a certain temperature all are liquid, but as we cool this "unified" liquid one by one they will condense out as solids. First, one will condense leaving the other two as a liquid mixture, then the second and the third. This is, in a sense, what happened to the forces of nature as the universe cooled. It is important to note, though, there were no physical changes in the fields as there would be in the case of the condensing liquids.

Now back to the X particle. We are dealing with the same interaction here as we are with the W particle, but the energy is now much greater—10^{15} GeV. This is the temperature at which the electromagnetic and strong forces were unified. Above it the X particle had no mass, but as the temperature cooled through it they gained mass through spontaneous symmetry breaking. At the same time the strong force froze out. Thus, just as the W's ate Higgs particles to gain weight, the X particle also ate what we call supermassive Higgs particles and gained considerable weight. This means that above 10^{15} GeV we had one family of particles—the combined lepton–quark family, which we now refer to as leptoquarks. Also we had one gauge particle of zero mass; the photon, the gluon, and the W's were all indistinguishable—they were all the same particle. The universe was obviously much simpler than it is now. Furthermore, the strengths of the fields, with the exception of gravity, were all the same—they were, in effect, the same force. If we made a plot of the coupling constants of the strong and electroweak fields, we would see that they move together with increasing energy (and

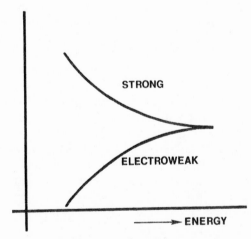

The merging of the strong and electroweak interactions at high energies.

corresponding increasing temperature) and finally at 10^{15} GeV they merge.

But what if the proton does decay? What are the implications? The most important consequences are, no doubt, in cosmology. If the proton decays to, say, a positron and a π^0, the π^0 will eventually decay to photons and the positron will meet with an electron and annihilate to photons. In short, all the matter of the universe will—in an incredibly long time—decay to radiation. There will be nothing in the universe but radiation! We could say the universe began as radiation (or at least was mostly radiation in the radiation era) and it will end, assuming it is open, as radiation. There will be no matter whatsoever left—a strange fate.

For experimental physicists, GUTs have another unpleasant aspect. We saw that the electroweak unification takes place at about 100 GeV. This is the approximate upper limit of our current accelerators; the next interesting energy is 10^{15} GeV—an energy we are not likely to attain with accelerators. This means that few things of interest will probably occur as we build ever larger accelerators—a rather bleak prospect.

SUPERGRAVITY AND SUPERSTRINGS

We have seen how the electromagnetic and strong interactions are unified in a theory we refer to as GUT. But this still leaves one field out of our scheme, namely gravity. If we are to have a truly unified theory of all the fields of nature, gravity must be included. So far, though, this has proven very difficult, mainly because the theory that describes gravity—general relativity—is a geometric theory and not a quantum theory. Although many scientists are hard at work trying to quantize it, it has not yet yielded. We can, of course, visualize the quantized version; like our other field theories, it would have to have a gauge particle to propagate the field. In the case of gravity, this particle is called the graviton. Thus, when two masses approach one another, there is a transfer of gravitons between them. Actually, since gravity is a long-range force—in theory it extends to infinity—there is a continuous transfer of gravitons between all objects in the universe, although, of course, when the objects are far from one another the number transferred is small.

A technique attempting to bring gravity into the fold that has been getting considerable attention lately is called supergravity. It is based on group theory, and the symmetry associated with it is usually called supersymmetry. To understand the main function of supergravity, we must remind ourselves of the nature of particles. Basically, aside from particular types, there are two fundamental kinds of particles in the universe: what we call matter particles such as electrons and protons, and exchange or gauge particles such as the photon and W particles. They are distinguished from one another by their spin: all exchange particles (called bosons) have integral spins, and all matter particles (called fermions) have half-integral spins ($\frac{1}{2}$, $\frac{3}{2}$, and so on).

What supergravity does is change fermions into bosons and vice versa. Just as we have a nucleon in isospin that can be made into a proton or a neutron by turning an imaginary dial, so can we imagine a superparticle in supergravity that is a fermion if the fermion arrow is up, or a boson if the boson arrow is up. In short, fermions and bosons are unified in supergravity; one can be changed into the other just as quarks can be changed into leptons in GUTs. This, of course, gives us the final step in the unification we have been seeking. In GUTs the early universe was simplified to two fundamental particles: a boson and a fermion. Now we have a way of transferring one of these into the other. This means that at the very earliest times the universe must have been exceedingly simple—perhaps consisting of a single type of particle. This would have happened at a temperature of about 10^{19} GeV, approximately 10^{-43} second after the big bang. Before this time, all four forces of nature were together as one unified force—with only one type of particle.

This is, in effect, the way the theory is believed to work, but the details and implications of the unification are far from worked out. Supergravity and its variations show considerable promise but there are still problems. In the simplest form of supergravity, the graviton is the only exchange particle; at higher energies, however, another particle called the gravitino is encountered. So far it has never been observed. These are the only two particles in the simplest form of the theory—in obvious disagreement with nature. But as we mentioned earlier supergravity can change an integral spin particle into a half-integral spin one. The graviton has spin 2, the gravitino spin $\frac{3}{2}$, and via supergravity transformations particles of spin 1 and $\frac{1}{2}$ can also be generated. The versions of supergravity in which these other particles appear are called extended supergravity. Many different kinds of particles are predicted in these extensions. In fact, to each known type of particle there is a super-partner; corresponding to the electron, for example, there is the selectron and to the photon there is the photino.

The addition of all these particles has helped in one respect: it seems to be giving us an understanding of renormalization. For years scientists have been getting around the difficulties of infinities in their theories by subtracting them out—in essence, by sweeping them under the rug. The method they used worked, but they were not quite sure why. With supergravity it seems as if we may be able to get around renormalization. Crudely speaking, it turns out that for each infinity in the theory that is caused by a boson there is an infinity of the opposite sign caused by a fermion, and they cancel one another.

Despite the promise of explaining renormalization, the theory does have difficulties. The major one is all the particles that are predicted by it—selectrons, winos, and so on. They have never been found in nature. Scientists, however, have an argument for this: They say that they might have been generated with so much mass that we have not yet been able to observe them. But when we get larger accelerators we will be able to.

Some of the difficulties of supergravity were alleviated recently when it was combined with another theory—the Klein–

Kaluza theory. We talked about it earlier. This theory was originally put forward by Kaluza in 1921. Kaluza extended general relativity to include electromagnetic theory by adding a dimension to it. But there were problems explaining this extra dimension. An explanation came a few years later from the Swedish physicist Oscar Klein; he told us that this extra dimension was everywhere, but it was rolled up in a little loop so tightly that we were unable to see it. According to his calculations, the radius of the loop was 10^{-33} centimeter. This is more than a billion billion times smaller than the nucleus of an atom.

Despite the apparent unification it seemed to bring, there was little interest in the Klein–Kaluza theory for many years. Then in the 1970s scientists began looking at it again, wondering if it might be of help in overcoming some of the difficulties of present-day theories. A modern version of the theory with 11 dimensions was soon devised; and again it was assumed that all but four of them were rolled up in tiny balls so that we could not see them. Scientists believed, in fact, that they acquired this property as a result of the big bang explosion.

So how does all this relate to supergravity? The relationship, it turns out, is with a version of supergravity called $N = 8$ supergravity. (The numeral refers to the number of steps that connect particles with different spins.) The major difficulty in finding a relation between supergravity and Klein–Kaluza theory was that supergravity was formulated in four dimensions and Klein–Kaluza theory was formulated in 11 dimensions. But then someone decided to look at supergravity in 11 dimensions and lo and behold there was a tremendous simplification (it became an $N = 1$ theory). With both theories now in 11 dimensions it was relatively easy to bring them together.

But again when the details of the resulting theory were worked out there were problems. The extra seven dimensions had to be rolled in a little ball to escape detection, and this affected the remaining four—the ones we see in our world. They also got compacted. Scientists were not ready to give up, however. The theory was tantalizingly close to solving many of the outstanding problems of physics, and almost everyone was sure we were on the right track.

What about a generalization of this theory?—perhaps the systematic addition of fields to it. The idea was tried, and sure enough, it worked. The resulting theory, called superstring theory, is believed by some to be the greatest breakthrough since general relativity. The "strings" idea came from the theory of strong interactions. Earlier we talked about quarks being held in their bags by strings. In our new theory, however, the strings are quite different. Work on the new theory was begun in 1979 by John Schwartz of Caltech and Michael Green of Queen Mary College in London. Few scientists were interested in the theory, however, until 1984 when Schwartz and Green showed that it could do many things other theories could not.

But what exactly are strings? They can best be thought of as one-dimensional bits of energy—like the strings of our experience, but only a billion billion trillionth of a centimeter long. They can be either closed like an elastic band, or open. They can interact with one another—two or more can coalesce and one can split in two. They can rotate and they can vibrate; in fact, it is this property that allows us to represent every type of particle we see in the universe. Each of these particles just has a different rotation or vibration. And again, as in Klein–Kaluza theory they are assumed to exist in a world of more than four dimensions—ten this time, six of which are rolled up and unobservable to us.

A modification of the theory, put forward in 1985 by David Gross and several colleagues of Princeton, has several new and interesting features. They found, for example, that it predicted four basic forces of nature, all of which went to one force in the early universe. Furthermore, they found they could literally predict all particles that we now observe. And perhaps best of all the theory was a geometrical one. The particles and forces were described geometrically in terms of the shape and vibrations of the strings. But general relativity is a geometric theory. Could this theory and general relativity be brought together? Work is now in progress to see if this is possible.

TWISTORS AND HEAVEN (H SPACE)

There are other approaches to the problem of joining gravity with the other fields of nature. Two of the better known are the twistor theory of Roger Penrose and the H space approach of E. T. Newman. Both of these theories take us from the world of real numbers into the world of complex numbers. A complex number is a number that consists of two parts: a real part and an imaginary part. Imaginary numbers play an important role in mathematics; certain equations could not be solved without them. Whenever we attempt to find the square root of a negative number, we find—assuming we are using real numbers—that we cannot. We must introduce what we refer to as i, the imaginary unit; it is the square root of -1. A complex number can be written as $a + ib$, where a is the real part and b the imaginary part. The introduction of complex numbers vastly extends the domain of mathematics, and has proven to be particularly useful.

Let us begin with Newman's theory. To understand it we should look back at the four basic types of black holes. They are as follows:

- S—Schwarzschild (nonspinning, noncharged)
- K—Kerr (spinning, noncharged)
- RN—Reissner–Nordström (charged)
- KN—Kerr–Newman (spinning, charged)

As of the mid-1960s solutions for the first three types had been found using general relativity. But no one had found a solution to the fourth. Then in 1965 Newman and Janis discovered an interesting relation between the first two. A simple transformation could be applied to the Schwarzschild solution that led to the Kerr solution. Soon afterwards, Newman and his students noticed that if the same transformation were applied to the RN black hole it would give a solution to the KN black hole. The relationship between the two pairs was the same.

Newman was immediately able to see why it had been so difficult to obtain a solution directly. The transformation took the RN solution, which was entirely real—in other words, in terms of only real numbers—into the world of complex numbers, namely the complex domain. Einstein's theory had never been solved in the complex domain; it was a real number theory.

Newman decided as a result of this to set up Einstein's theory in terms of complex numbers, and explain why the transformation worked. This led him to introduce a complex space he calls H space (a pun on heaven). Considerable progress has been made toward the realization of his goal and he now feels he understands why the transformation works.

Closely related to H space is Roger Penrose's twistor space. Penrose was dissatisfied in the dichotomy between quantum theory and general relativity. While quantum theory was based on complex numbers, general relativity was based on real numbers. He wanted to unify the two theories by bringing general relativity into the complex domain. His twistor space has eight dimensions, one real and one imaginary for each of the four dimensions of ordinary space-time. He calls the components in this twistor space twistors.

What are these twistors? That is a little difficult to answer. They are not exactly particles, or points in space, but a kind of combination of both. According to Penrose, all the particles of nature are made up of twistors. Gauge particles are made up of two twistors, particles like electrons of two twistors, and heavy particles of three twistors. But this is not their only function. Penrose has always disliked the concept of empty space—a background on which particles move. He showed that his twistors also make up the space itself. A point in space is, in effect, a collection of twistors.

As we have just seen, many new techniques have been

brought to bear on the problem of unification. We have seen how a new group has been used to give a tentative unification of electroweakdynamics and the strong interactions (GUTs). We have seen how supergravity and superstring theory are attempting to bring gravity into the fold. Other theories such as twistor theory also hope to accomplish this. But so far no theory has been successful.

CHAPTER 11

Epilogue

Einstein's dream of a unified theory of the universe has not yet been achieved, but with the successes of the last few years it appears as if we are at last within sight of the goal. Most scientists, I am sure, would not care to hazard a guess as to when this will be achieved, but they are now fairly certain it will eventually happen.

Our present goal is, however, different from Einstein's. Einstein's attempt was, most will agree, ahead of its time; there was still much to be learned. Scientists at that time knew nothing about the large number of elementary particle types, nothing about symmetry in nature, nothing about gauge theory, and very little about the big bang that started it all.

Because Einstein had a deep belief in causality, he wanted a theory that was strictly causal, i.e., for every effect there had to be a cause. He also wanted his theory to unify only electromagnetism and gravity; in this respect it was quite different from the modern endeavors. Furthermore, he wanted the particles of physics—the few that were then known—to emerge as solutions of the equations of his new theory. And perhaps most important of all, he wanted quantum theory to emerge as a first approximation to his theory.

The theory that scientists are striving for today goes far beyond Einstein's dream. They want a theory that links quantum mechanics and general relativity, and one that encompasses the forces and elementary particles of nature; in short, they want a theory of everything (TOE). Furthermore, this theory should have a mathematical beauty and basic simplicity about it. We have just seen that considerable progress has been made toward such a theory. Major clues, it seems, reside in our theories of the early universe. The universe, at that stage, may have been much simpler than it is today. There may have been a single force—a *superforce*—that gave us all the present forces and particles of nature.

The problems that remain in perfecting the theory are likely to be difficult ones. But because scientists have an insatiable curiosity about nature, they will persevere. There is, however, a difficulty: with the introduction of strange new particles, strings, twisted, stretched, and foamy space-time, 11 dimensions, and so on, the universe has become increasingly difficult to comprehend and visualize. Prejudices get in the way, but scientists have to keep an open mind. They must be able to stretch their minds to regions that have never been explored before. Fresh, novel ideas are essential if progress is to be made.

Strangely, though, as the universe has receded from our world of senses a deep remarkable harmony has become evident. There is order and unity that we never dreamed of. Each breakthrough seems to uncover new coherence, new order, new unity.

This unity even seems to extend to life itself. Stephen Hawking has looked at what might happen after the "big crunch" of the universe (assuming it is closed). He has found that if there is a bounce and a new universe is produced, the fundamental constants of nature (e.g., mass and charge of the electron) will change. Brandon Carter of Cambridge University has followed up on this development and has arrived at some amazing results. He showed that life could not exist in a universe with different fundamental constants. Change them slightly in one direction and giant stars would not exist, and if they did not, where would our elements come from? Change them slightly in the other direction and only small red stars would exist; there would be no stars like our sun, and therefore no good candidates for life. It almost seems as if life is somehow tuned to the present fundamental constants of the universe. Change them slightly and you tune out life.

THE FUTURE

At this point we might ask: Where do we go from here? One of the things that is definitely going to help, assuming that it is built, is the Superconducting Super Collider (SSC). It will take more than 10 years to build, but will be far larger and much more powerful than anything in existence. It may, for example, allow us to see the Higgs particle, which would be a further verification of GUTs. Ten thousand superconducting magnets will be used to bend the beam into a circular path, so it will have considerable girth. This means it will likely have to be located on a large flat area, perhaps a desert. So far we do not know where this will be, but almost every state in the Union is vying for it.

Many scientists believe that supergravity and superstring theory are the wave of the future and will lead the way over the next few years. Supergravity is a remarkable theory; it predicts many of the particles we presently assume exist, such as quarks, but it also predicts particles such as selectrons and winos that we do not see. Perhaps, though, with the SSC we will find them. Superstring theory, on the other hand, seems to predict most of the particles we observe. Freeman Dyson has recently said, "Supergravity is in my opinion the only extension of Einstein's theory that enhances, rather than diminishes the beauty and symmetry of the theory."

GUTs have also had a number of successes in recent years; we have recently detected the W and Z particles and seen evidence for the existence of quarks. This has been helpful but problems remain. Some of the predictions it makes have not yet been verified. For example:

1. It predicts the proton should decay in about 10^{31} years, and there are indications its lifetime is longer than this.

2. It predicts neutrinos should have mass and should oscillate in type. This has not been verified.
3. It predicts the magnetic monopole, which has not been found.
4. It predicts the Higgs particle, which has not been found.
5. We have never directly seen some of the particles GUTs says should exist, for example quarks and gluons. (This may of course not be a flaw in that the theory says we will never observe them.)

In relation to the overall universe, we also have important questions yet to be answered:

1. Is the universe open or closed?
2. How did the universe come into being?
3. What was here before the big bang?

Recent evidence seems to indicate the universe is close to the borderline between being open and closed but we are still not certain. The second question has generally been ignored by scientists until recently; they felt it was something we just could not answer, something that was forever beyond their reach. But with inflation theory, scientists feel they may be able to answer it. The universe may have started from nothing—a tremendous amount of energy may not have been needed. It may, in a sense, have been self-generating—once it got started it generated its own energy. Finally, in regard to what came before the big bang, we are definitely out on a limb here.

THE NEW EINSTEIN

We might ask ourselves what Einstein would think of our progress. We know, of course, that he never liked quantum theory but Peter van Nieuwenhuisen, one of the originators of supergravity, has said that if Einstein had studied anticommuting numbers (numbers a,b such that $ab + ba = 0$) he might have invented supergravity. So perhaps he might like it.

Will a new Einstein come along and put it all together, or will it gradually be put together into a final theory by many people—each making only a small contribution? In the past it seemed as if great strides have usually been made by individuals—individuals like Einstein. And somehow this seems likely to be the case again. What is important is to be able to look at old ideas in a new way. As some have said, "You need crazy ideas—crazy enough to work." And the new Einstein will certainly need crazy new ideas.

Further Reading

GENERAL

Ferris, Timothy, *The Red Limit* (New York: Morrow, 1983).
Pagels, Heinz, *The Cosmic Code* (New York: Simon and Schuster, 1983).
Parker, Barry, *Concepts of the Cosmos* (San Diego: Harcourt, Brace, Jovanovich, 1984).

GENERAL RELATIVITY AND COSMOLOGY

Asimov, Isaac, *The Collapsing Universe* (New York: Simon and Schuster, 1977).
Kaufmann, William, *Black Holes and Warped Spacetime* (San Francisco: Freeman, 1979).
Shipman, Harry, *Black Holes, Quasars and the Universe* (Boston: Houghton-Mifflin, 1980).
Silk, Joseph, *The Big Bang* (San Francisco: Freeman, 1980).
Trefil, James, *The Moment of Creation* (New York: Scribners, 1983).
Weinberg, Steven, *The First Three Minutes* (New York: Basic Books, 1977).

PARTICLE PHYSICS

Calder, Nigel, *The Key to the Universe* (New York: Viking, 1977).

Feinberg, Gerald, *What is the World Made of?* (New York: Doubleday, 1977).

Trefil, James, *From Atoms to Quarks* (New York: Scribners, 1980).

Many articles can also be found in the following magazines:

Astronomy
Sky and Telescope
Mercury
Science Digest
Discover
Science 86
Smithsonian
Scientific American

Index